WIND / SOLAR ENERGY

for
Radiocommunications and Low-Power
Electronic / Electric Applications

by
Edward M. Noll

Howard W. Sams & Co., Inc.
4300 WEST 62ND ST. INDIANAPOLIS, INDIANA 46268 USA

Preface

Wind/Solar Energy for Radiocommunications and Low-Power Electronic/Electric Applications is an introduction to the practical use of sunlight in the construction of solar power supplies and the conversion of wind energy to electricity. The objectives of the book are to give you an insight into the techniques used in converting light energy and wind energy to electrical power. Several practical supplies are described in detail and permit you to gain that initial experience. More elaborate yet moderate installations are discussed, and methods of making your house or small business more self-sufficient are described in terms of electrical needs.

Chapters 1 and 2 discuss the basic principles of both methods of energy conversion. You will learn circuits and techniques. Chapter 3 describes the all-important electrical storage unit of a solar or wind supply. Secondary batteries are covered in detail along with descriptions of inverters which are used to convert dc to ac electricity. The electrical vehicle, which is a natural receptor for power generated by solar energy, is covered briefly. Chapters 4 and 5 describe practical systems. The appendix lists addresses of suppliers and sources of material and information concerning solar power and related subjects.

A special thanks is given to Jim Fisk, editor of *Ham Radio Magazine,* for his encouragement and permission to reuse material that first appeared in article form as a part of my regular column "Circuits and Techniques" in *Ham Radio Magazine.*

EDWARD M. NOLL

Contents

CHAPTER 1

CHAPTER 2

CHAPTER 3

Solar Energy and Light Converters

Praised be my Lord God with all his creatures; and especially our brother the sun who brings us the day, and who brings us the light; fair is he, and shining with a very great splendor:

. . . .

Praised be my Lord for our brother the wind, and for air and cloud, calms and all weather, by the which thou upholdest in life all creatures.   *Francis of Assisi*

The sun blesses us with warmth and light, each a nonpolluting form of abundant energy. By indirect means the sun begets the wind, a purifier extraordinary and conveyer of tremendous energy. The ever-shifting differential heating of the earth's surface by the sun launches the wind. It blows over the wind-swept plains, races above the mountain tops, and dances along the sea cliffs and out over the oceans.

How much converted electrical energy does this involve? Capacity estimates vary from 20 billion kilowatts for choice sites to an 80-trillion-kilowatt capability for the entire northern hemisphere.

The sun is a magnificent nuclear furnace and is pouring energy upon us in vast quantities. Thus far we have chosen not to use it fully or wisely. Rather we derive most of our energy from limited fossil fuels and the polluting fabrications of man.

Light photons from the sun bombard us continuously with energy on both bright and overcast days. This light can be converted to electrical energy by photovoltaic cells. How much electrical energy does this involve? Even at an average efficiency of only 10%, a cell-bed of 1000 square feet could develop a significant percentage of the total electrical needs of the average home. Such large installations are costly today. But the hope for tomorrow is most encouraging.

Today there is no technical reason why farms using both sun and wind energies cannot be made self-sufficient in terms of heating and electricity. This also applies to country, small town, and many suburban homes. There are condominiums under construction or being planned that will be entirely self-sufficient in terms of energy needs. In more crowded areas, at least some proportion of self-sufficiency can be obtained, especially for several days of emergency or blackout conditions.

This is a very practical book. The emphasis is on practical systems for generating small quantities of electricity by light energy and wind-energy conversion. Commercial and home-brew systems for generating electricity for the home or small business are also discussed. Special emphasis, however, is given to the use of wind/solar systems that supply electrical power for electronic equipment.

THAT CRAZY OLD SUN

How much energy is poured upon us by this single thermonuclear generator, located a safe distance away? Solar cells spread over 1% of the Sahara desert could produce the total electrical energy consumed on the earth today. At high noon, the solar energy impinging on 5000 square miles of the earth's surface matches the peak capacity of the earth's existing power plants. There is indeed enough solar energy to go around.

A solar panel installation that occupies 1000 square feet of your roof, suitably tilted and oriented toward the south, can deliver 60–80% of the power requirements of an average home. Double this area and you have it made.

Today the cost is prohibitive. But the energy is there and the system does work. Practical, low-cost installations are now a matter of engineering, experimentation, and demand.

The elements that compose a photovoltaic cell are available. Silicon is one of the most common elements on this earth. It is estimated that in less than ten years the refining technique that produces silicon cells will be improved to the extent that costs will fall from $5 a watt to less than 50¢. Special efforts are being directed toward improvement in the conversion efficiency between photon energy and electrical energy. Such efficiency is now in the 7–10% range. An increase in conversion efficiency would permit size reduction and less roof area would be needed for a given power output.

Funds are needed for experimentation and the development of large pilot installations. Since the research and development money comes from the government and foundations, as well as from corporate contributions, the factor of demand becomes important. This means education, too, is important, and the public must be convinced that this is worthwhile.

Energy from the sun arrives in the form of heat and light photons. The energy of the light photons can be converted to electrical energy. Heat from the sun can be used to heat and air-condition a dwelling. In fact, a photovoltaic solar panel can also be employed as a solar heat collector, serving the dual function of supplying electrical energy as well as heat for the solar heating system of the dwelling.

The electrical powering of an electronics shop or radio ham station (up to its 1-kilowatt power level) is very feasible today. Of course, the cost is higher than operating a comparable power mains supply. However, it is an important area of experimentation with ultimate cost diminishing sharply, relative to the rising cost of commercial electricity.

Such a solar power supply affords emergency power during commercial blackouts and brownouts. It can be used to maintain charges on batteries used for mobile application, for portable operations, or at any site at which there is no commercial power, or when the use of commercial power is inconvenient.

The wind is a secondary source of energy freed by the sun. At present costs, a combination of wind generator and solar panel provides an excellent solar energy alternative. The wind generator can be used to charge a group of husky batteries. The solar panel can be used for low-powered equipment and to provide a trickle charge for the high-powered batteries of a solar power supply. Details on such installations are given in succeeding chapters. The author's electronics bench and radio ham station are being powered largely on solar energy.

BASIC SYSTEMS

Light-Energy Converter

The basic light-energy converter, shown in Fig. 1-1, makes a conversion between photons of light energy and dc electricity. This dc electricity can be used directly, or it can be used indirectly to charge batteries or activate some other type of energy-storing method.

First let us put aside the notion that solar cells are only effective in bright sunlight and that only in the southwest, with its endless clear days and bright sun, are solar-powered devices

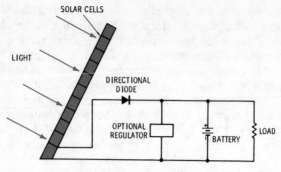

Fig. 1-1. Basic solar power supply.

feasible. Actually, solar cells work quite efficiently in areas with a lower number of clear days per year. They also function well during bright, overcast days. Even dark days produce some output, although at a much lower level. The secret is to match the installation with local climatic conditions. The number of cells required for a specific application depends directly on the average weather conditions at the site. This simply tells us that in the east and northeast, with their more generous portion of cloudy, rainy days, more solar cells are needed per given power demand than would be required in the sunny south and southwest.

However, for each part of the country, average solar energy levels have been measured for many years. Using this data you can select the proper number of cells required and add a few additional ones for good measure to accommodate a long sequence of cloudy winter days.

To make efficient use of solar light, the solar panel and its cells must be tilted in a southerly direction. The average tilt angle corresponds with the latitude of the site (number of degrees north of the equator). If your solar panel is mounted where it is readily accessible, you can make minor adjustments in this angle in the spring and fall to compensate for the somewhat different path taken by the sun across the heavens between dawn and dusk. (More exactly, in an astronomical sense, compensation is made for the changing tilt angle of the rotating earth, the sun being in a fixed position.)

A solar panel or light-energy converter can be designed to deliver a certain amount of power in daylight. However, any solar energy system, to be worthy, should also be capable of taking care of nighttime needs. In terms of electrical power energy, this can be handled by chargeable batteries. Thus, a 24-hour, all-year-round system requires both the light-energy converter and the energy-storing battery system. Such a system can be combined into a completely self-sustaining and self-maintaining, unattended installation.

Later chapters detail how a successful solar power system can be planned to meet the power requirements and the average weather conditions of a specific location. Voltage and peak power requirements must also be met. Consideration must be given to the average power demand over a 24-hour period and then over a long-term solar-year basis. The solar panel of Fig.

Courtesy Solarex

Fig. 1-2. A 200-watt solar array.

1-2 has a 200-watt peak capability. This means that in optimum light it will supply 200 watts of energy to the energy-storing system. As shown in Fig. 1-1, a protective diode system is often included. Such a precaution prevents the discharge of the energy-storage device when the impinging light is not strong enough to permit charge activity. Voltage or current regulator circuits can be included in the output to match the specific needs of the energy-storing plan.

Wind Generator

The ever-shifting differential heating of the earth by the sun's warmth causes the motion of the winds. The conversion of the tremendous force of wind motion to electricity is another form of nonpolluting solar energy. The world is struggling to meet the demand for electrical energy and is consuming fossil-fuel sources at a devastating rate while only lackluster impetus is given to the use of the vast quantity of nonpolluting energy from sun and wind.

A basic function diagram is shown in Fig. 1-3. The initial step in the conversion of wind energy to mechanical motion can be accomplished by a variety of wind blades and rotors. Such blades and rotators are described in succeeding chapters.

Electrical generators of various types make the conversion from mechanical motion to electricity. Sometimes an intervening belt and/or gear system is used to make an appropriate conversion between the rotational speed of the wind-driven rotor and a suitable speed of rotation for the generator. Other generators are driven directly by the wind-driven rotor. Either dc or ac generators can be employed with various voltage and power capabilities as required by the system. Popular values are 12-, 24-, 36-, and 110-volts dc.

There are periods of calm and low wind velocity. Generators become inactive. Electrical energy must be stored by batteries or other energy-storing methods. Presently the secondary battery is popular and serves as a reservoir of electrical energy.

Wind electrical-generating systems must be planned according to the average wind-velocity conditions of an area, with an added safety factor that would accommodate a reasonably long wind-quiet period. Most practical systems today use the wind generator to charge batteries. Usually this dc voltage is then distributed from the battery to wherever it is needed. Direct-current-operated bulbs, motors, and other electrical devices are readily available. For those electrical devices that require 110-volts ac, it is necessary to insert an intervening inverter. Solid-state inverters with ratings between 50 and 75 watts, and as high as 5 kilowatts, are available.

The planning of a wind-generator installation is related directly to the local wind conditions. This information has also been compiled over a number of years. At the present state of the science, a rule of thumb is that an average yearly wind velocity of 9–10 mph is adequate for the installation of a practical wind-generating system.

The real energy-producing winds fall in the 15–25 mph range. When such levels are maintained for an average two days out of seven throughout the year, you are in a practical situation for a wind-generating system. If you are on a site that

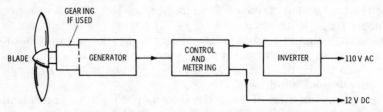

Fig. 1-3. Basic plan of wind generating system.

betters these values, you will have a bonus in extra power. The installation of Fig. 1-4 in East Holden, Maine supplies the electrical needs of a small business. Presently the 2-kW generator on the self-supporting tower makes the office independent of the electrical mains. In the background is a newer 6-kW model under test by this small company in the wind power business.

LIGHT MEASUREMENT

The Langley is considered a unit of solar radiation. It is equivalent to 1 calorie of radiation energy per square centimeter. Of course, this intensity varies with geographical location, daily and seasonally. Also, clouds, dust, and pollution decrease the amount of solar radiation reaching the earth.

Customarily, our meterological stations report solar radiation in terms of the total Langleys impinging on a horizontal surface at ground level. However, in terms of the wise utilization of solar radiation, a solar panel, shown in Fig. 1-1, would be tilted at a more favorable angle toward the sun, causing the sun's rays to strike the panel at a more perpendicular angle as the earth revolves on its axis relative to the sun's position. Actually, a radiation of 1 Langley per minute is an acceptable average for such a tilted panel in sunlight. This average can be increased by about 40% if the sun is tracked across the heavens. The combination of tilting and tracking provides the benefits of maximum radiation.

Let us consider the practical significance of an average radiation of 1 Langley/minute. One kilowatt of heat from the sun's radiation corresponds to 14,300 calories/minute. If the measured solar radiation is 1 Langley, an area of 1 centimeter squared produces 7×10^{-5} kilowatt. To increase this level to 1 kilowatt requires an increase in area of 14,300 $(1 \div 7 \times 10^{-5})$ times. Therefore, the surface would have to be 14,300 centimeters squared, or 1.43 meters squared. This corresponds to an area of approximately 15.4 square feet.

An important consideration is how efficiently this solar power can be converted to electricity. Assuming a conversion efficiency of 6%, the derived electrical power would be 60 watts (1000×0.06). This amounts to approximately 4 peak watts/sq ft $(60 \div 15)$.

It has become acceptable to use a standard test condition (STC) in solar-power engineering. Standard value is 100 milliwatts per centimeter squared solar intensity. This value is based on the fact that the solar intensity at noon on a clear day

Fig. 1-4. 2-kW and 6-kW wind generators.

is about 100 milliwatts per centimeter squared at 25°C (77°F). For a typical solar cell operating under this intensity, the short-circuit current is approximately 25–30 milliamperes per square centimeter, and the open-circuit voltage is 0.55–0.6 volt (Fig. 1-5). At higher temperatures the current declines, and this fact must be considered in the derating of a practical solar panel.

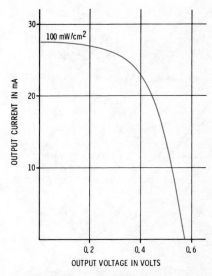

Fig. 1-5. Output characteristics for standard solar intensity.

A practical solar module, approximately 1 foot square, will deliver 300 milliamperes under a light intensity of 100 milliwatts per centimeter squared, even at an operating temperature of 140°F. Output is less when intensity level declines.

In determining the quantity of electricity made available in kilowatt-hours, averaging figures are sometimes based on eight hours of sunlight. Often, a more conservative figure of 5 hours of sunlight per day is more practical. Thus, a solar panel delivering 60 peak watts for a period of 5 hours would supply 300 watt-hours (5 × 60) for storage. Averaged over a day this would correspond to a continuous drain of 12.5 (300 ÷ 24) watts. On a monthly basis, using a 30-day figure, the quantity of electricity made available would be 9 kWh (30 × 0.3).

Variations in solar energy must be considered in planning a solar power system. These involve weather conditions, the earth's rotation, and seasonal declinations. Solar measurements

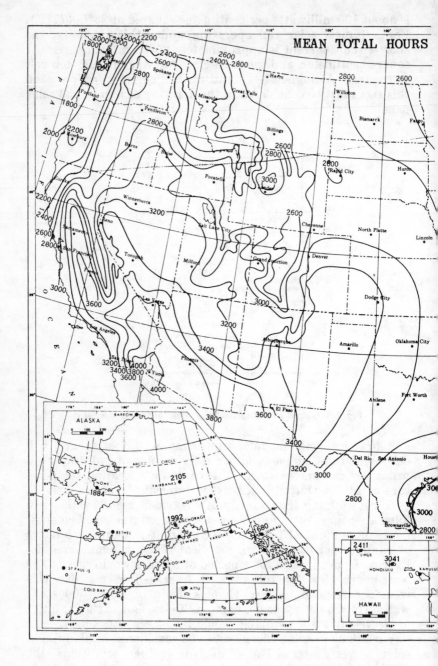

Fig. 1-6. The mean total

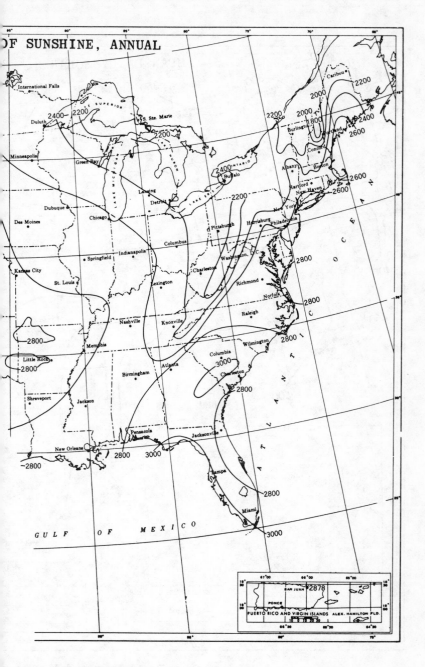

hours of sunshine—annual.

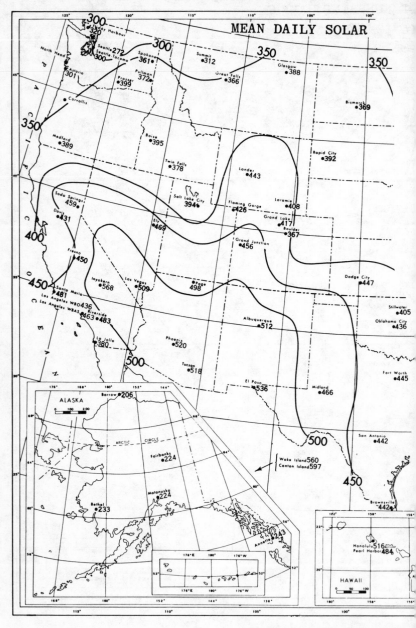

Fig. 1-7. The mean daily solar

18

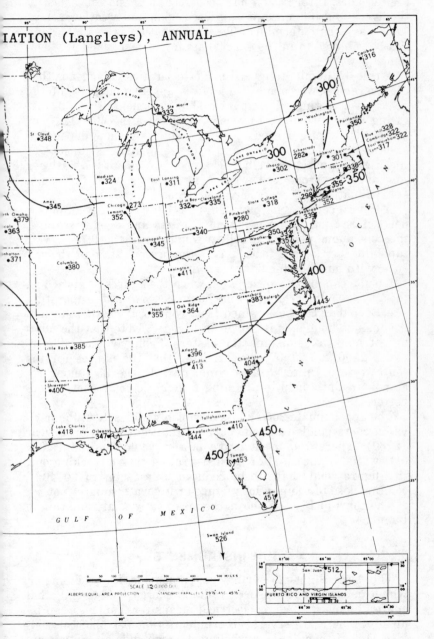

radiation in Langleys—annual.

have been made at various sites throughout the world for several decades. Information has been recorded, and consequently the averaging influence of daily, seasonal, and weather conditions has been determined. This information is available and can be referred to in the exact planning of a solar energy system.

A rule of thumb often used is a 5-to-1 ratio. This means that the average power made available is approximately $\frac{1}{5}$ of the peak power rating of the panel. Thus, a panel with a 60-watt rating would provide an average of 12.5 watts (60 ÷ 12.5) of power. The figure can be above or below this value depending upon geographical location and average weather conditions at the site. In a practical vein, this amounts to about 0.7 watt per square foot of solar cell area. A figure often used for rough approximations is 1 watt per square foot.

A typical commercial panel supplies 3.6 watts of peak power per square foot. Using averaging figures and energy storage, the continuous power capability would be in the 0.75–1 watt capability per square foot. This is the present state of the science, with much better figures on the way.

Results can also be anticipated when the figures are based on the average number of sunshine hours per year at the operating site. Such data can be obtained from your local weather station. A sunshine-hour map is shown in Fig. 1-6. Note that the number of hours of sunshine for eastern Pennsylvania falls between 2500 and 2700 per year. Note that the figures for the southwestern United States climb up into the 3500-hour region—a definite help in planning a gigantic solar energy converter.

The approximate 2600 hours of sunshine in eastern Pennsylvania corresponds to about 30% of the total hours in a year (8760). Therefore, the average power would be 1.08 (3.6 × 0.3) watts. In the case of a fixed site rather than a tracking one, this figure would have to be reduced by a factor of 30–40%, coming close to the previous figure of 0.7 watt/square foot.

The chart of Fig. 1-7 shows the mean daily solar radiation in Langleys.

SUN INSTRUMENTS

The basic light-measuring instrument is the radiometer, which detects and measures radiant energy, converting that energy by mechanical or electrical means for display on an indicator. There are several instruments that perform this service. One of them is the pyrheliometer, which is an instrument that measures the intensity of the direct solar beam. The pyranometer, shown in Fig. 1-8, measures the total radiation from both sun and sky. It is used widely to measure solar energy. The term *photometer* applies to instruments used for measuring total radiation.

For many years, costly thermopile pyrheliometers and pyranometers have been used to measure solar radiation. These have been replaced by accurate and less costly radiometers using electronic devices, mainly the silicon photovoltaic cell.

Courtesy Spectrolab

Fig. 1-8. Solar pyranometer.

The silicon cell is an excellent device for a radiometer because there is a linear relationship between its short-circuit current and the intensity of the solar radiation falling on the cell. As is typical of semiconductor devices, there is a rise in this short-circuit current with temperature. Proper compensation in the circuit includes a resistor with a negative temperature coefficient. For more accurate performance, a thermistor can be used. A thermistor in conjunction with Manganin wire can be used to establish any desired negative temperature-resistance characteristic.

The Matrix, Inc., Sol-A-Meter, seen in Fig. 1-9, is a weatherproofed solar radiometer. It employs a silicon photovoltaic cell as the sensor, and its spectral response extends from 0.35 to 1.15 micrometers (microns). In manufacture the instrument is calibrated with a thermopile reference radiometer. In addition, it includes solid-state integrated circuits and provides a direct readout on a digital display. Although the instrument is battery operated, it uses a nickel-cadmium type, which is recharged by a built-in solar charger.

Remote sensors and readouts are available. The instrument can be equipped with an inkless type of recorder. The latter includes either a 6- or 12-volt dc or 115-volt 60-Hertz ac motor. A photocell relay can also be included which will turn off the recorder at sunset and turn it back on just before sunrise.

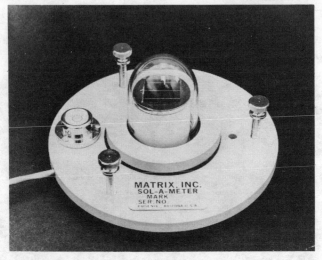

(A) Sunlight sensor.

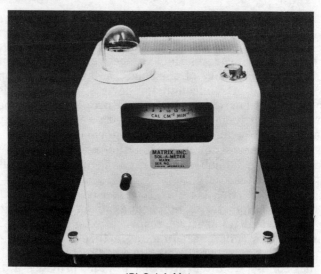

(B) Sol-A-Meter.

Courtesy Matrix Inc.

Fig. 1-9. Sun meters.

A well-designed and calibrated solar panel can be used with an ampere-hour meter such as shown in Fig. 4-1. Such a meter permits one to keep day-by-day, month-by-month records of light converted to ampere-hours at a specific location.

SILICON CELL PRINCIPLES

The element silicon is basic to the most common photovoltaic cell. The characteristics that make one element differ from another are the grouping, number, and placement of electrons around the atomic nucleus. Electrons may occupy one or more energy bands, or shells. Shells are spaced a certain distance from the nucleus and contain a specific number of electrons of approximately the same energy level. Those electrons in orbit near the nucleus have a low energy level and are tightly held to the atomic structure. Electrons in the outer shell are of much higher energy content and are held less tightly.

The makeup of the silicon atom is demonstrated in Fig. 1-10. It consists of 2 electrons in the first shell (K), 8 electrons in the second shell (L), and 4 electrons in the outer shell (M). Customarily, a simplified version, shown in Fig. 1-10B, is used to show the nucleus and the four electrons of the outer shell. It is the outer, so-called valence electrons that determine the electrical characteristics of the particular element.

There are never more than eight electrons in an outer shell, but there may be fewer. When there is exactly eight, an element has a high stability because the valence electrons are bound tightly to the atom. Such an element serves as an excellent electrical insulator.

Fewer than eight valence electrons result in a less stable atom. Atoms having 5, 6, and 7 electrons tend to borrow additional electrons from other atoms. Those with 1, 2, or 3 valence electrons lose electrons to other atoms.

The interlocking of valence electrons among atoms produces stable molecules in a substance, forming crystalline formations of molecules. Silicon is an unusual element because there are four valence electrons in a balanced arrangement. Its binding is rather tight, and it takes on the characteristics of an insulator.

When highly purified silicon is produced, it is called intrinsic silicon. Although its conductivity is poor, its crystalline makeup is not a perfect insulator because the bonding can be broken up with high temperature, light photons, or electrical energy.

In the manufacture of semiconductor devices, very small and very accurate amounts of impurities are added to the intrinsic crystal. These impurities establish the polarization of the material and lower its resistivity by a specific amount, giving a very carefully regulated amount of electrical conduction. Such a crystal with the proper amount of impurity (doping) is called an extrinsic semiconductor. It has a conductivity that lies somewhere between the high conductivity of a conductor and the low conductivity of an insulator.

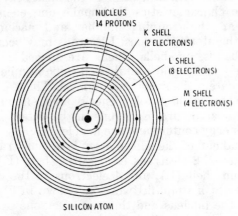

(A) Silicon atom showing energy bands.

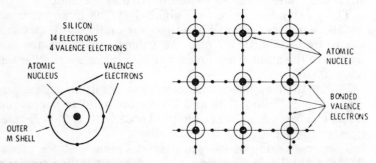

(B) Simplified drawing of the atom.   (C) Atoms arranged in a lattice.

Fig. 1-10. The silicon atom.

Commonly the impurities added to the intrinsic silicon crystal are of two types: one has only three electrons in the valence shell, and the other five electrons in the band. The makeup of the more common impurity elements is shown in Fig. 1-11. Elements with three electrons in the valence band of their atom, such as boron or gallium, are called acceptor materials and, when added to pure silicon as an impurity, produce p-type semiconductor material. Elements that have five valence electrons are phosphorous, arsenic, and antimony. When added to silicon, these donor atoms produce n-type material.

The type of doping of the silicon crystal determines whether it acts as a p-type or an n-type material. If the impurity has five valence electrons, the bonding of the material is as shown

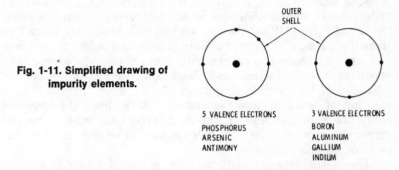

Fig. 1-11. Simplified drawing of impurity elements.

OUTER SHELL

5 VALENCE ELECTRONS
PHOSPHORUS
ARSENIC
ANTIMONY

3 VALENCE ELECTRONS
BORON
ALUMINUM
GALLIUM
INDIUM

in Fig. 1-12A. Note that at every position where there is an impurity atom, there is an extra electron in the outer orbit. Such a crystal is called n-type because the donor doping has produced extra electrons which can be moved as negatively charged particles. It is important to note that the semiconduc-

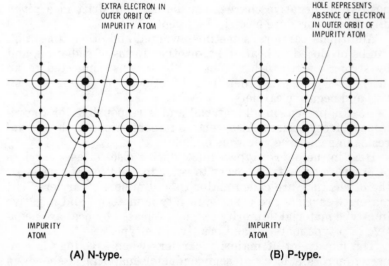

EXTRA ELECTRON IN
OUTER ORBIT OF
IMPURITY ATOM

HOLE REPRESENTS
ABSENCE OF ELECTRON
IN OUTER ORBIT OF
IMPURITY ATOM

IMPURITY
ATOM

IMPURITY
ATOM

(A) N-type.

(B) P-type.

Fig. 1-12. Impurity atom in silicon lattice structure.

tor material itself does not have an overall negative charge because the total number of electrons in the substance equals the total number of protons. However, it is said to have electron carriers which can be moved with the application of a suitable outside force.

If the doping element has three valence electrons, the bonding of the intrinsic atoms is such that there are electron vacancies, as shown in Fig. 1-12B. Such an empty charge position, at which there would otherwise be an electron in a complete bond, is called a hole. It constitutes an electron absence and can be considered as a positive particle. There is always a tendency for an electron from a neighboring atom to move into the empty position or hole. In this case, there is a hole left in the bonded makeup of the neighboring atom. As a result, there is a free motion of holes or positive charges throughout the material with the application of an external force. Current that results from the random motion of holes is said to be supported by the motion of hole carriers.

The actual conductivity of the material depends on the amount of doping. The higher the doping, the greater is the number of electron or hole carriers. Current in a semiconductor is usually spoken of as movement of positive or negative carriers instead of hole or electron carriers, respectively. In n-type semiconductor material, current is the result of the motion of negative charges (electrons), while in p-type material, current results from the motion of positive charges (holes). These positive or negative charges can be made to drift in a given direction to produce electrical current.

Additional carriers, sometimes wanted and other times not, can be produced by heat, light, or other forms of radiation, and by strong electric fields. When a battery is connected to an intrinsic-type crystal, there is a directional movement of positive and negative carriers.

Heat is present in all material and is responsible for the release of free carriers. Also, certain imperfections in the crystal result in some free electrons or holes.

Heat in the p- or n-type semiconductor also releases carriers of opposite polarity. In order to refer to one type of carrier or the other, they are called majority and minority carriers. The majority carriers are positive in p-type material and negative in n-type material. Minority carriers released by heat are negative in p-type material and positive in n-type.

The movement of majority carriers when a voltage is connected across a block of semiconductor material is shown in Fig. 1-13A. Note that in the n-type material of A, the negative

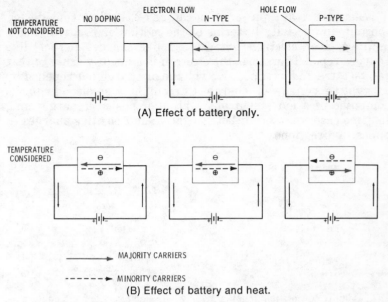

(A) Effect of battery only.

MAJORITY CARRIERS

- - - - - - ▶ MINORITY CARRIERS

(B) Effect of battery and heat.

Fig. 1-13. Charge flow in a semiconductor.

electron carriers produce a drift toward the positive terminal of the battery. The n-type material contains hole minority carriers, and these positive charges move toward the negative terminal of the battery, as shown in Fig. 1-13B. The minority carrier motion is the result of heat in the semiconductor material. Such carriers increase with a rise in temperature. This is an undesirable feature in semiconductor devices. In practice the number of minority carriers is relatively small as compared with the majority carriers.

In the p-type material, majority carriers are positive particles or holes. They move toward the negative terminal of the battery as shown in Fig. 13A. In the p-type material, the minority carriers are electrons, and these negative particles move toward the positive terminal of the battery.

The PN Junction

The semiconductor pn junction is a part of most solid-state devices—diodes, photovoltaic cells, bipolar transistors, field-effect transistors, integrated circuits, etc. The types and applications of such junctions seem endless because of the versatile manner in which characteristics can be controlled by shape, extent of doping, type of material, manner of activation, and other factors.

What occurs at a pn junction can be clarified by considering the junction activity in terms of the motion of positive (hole) and negative (electron) charges. Unlike charges attract; like charges repel. Therefore, the electron is a negative charge that can be attracted by a positive voltage or charge but repelled by a negative voltage or charge. Current in a semiconductor is composed of a movement of negative or positive charges and, in some cases, motion of both negative and positive charges in opposite directions.

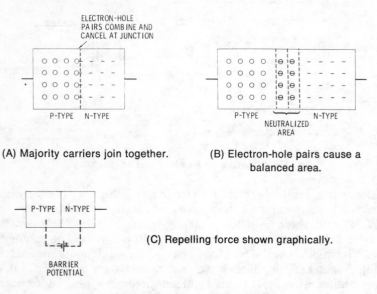

(A) Majority carriers join together.

(B) Electron-hole pairs cause a balanced area.

(C) Repelling force shown graphically.

Fig. 1-14. Forming the pn junction barrier.

Actually, when a pn junction is formed, and even without the application of a bias, the majority carriers near the junction attract each other. They cross the junction and cancel as shown in Fig. 1-14. This cancelling action of the mobile carriers in electron-hole pairing establishes a charge between the two types of semiconductor material. Since the majority carriers near the junction have cancelled, the semiconductor material near the junction has a charge that tends to hold the majority carriers away from the junction, as shown in Fig. 1-15. In effect, the electrons in the n-type material are repelled by the negative charge of the p-type material, and the holes of the p-type material are repelled by the positive charge in the n-type material. The majority carriers, therefore, maintain positions back from the junction. This repelling force is called

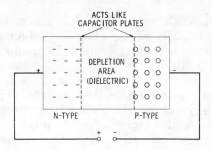

Fig. 1-15. The capacitive effect in a semiconductor.

a potential barrier. In the case of silicon, it amounts to a charge of approximately 0.5 volt. This barrier potential must be overcome before it is possible to move majority carriers freely (electrons or holes) across the junction.

An important fact about the pn junction is that the establishment of the potential barrier depends on the movement of electrons in one material and holes in the other. Electrons are bound or immobile in p-type material, and current depends on the movement of holes. In n-type material the holes are immobile, and current depends on the movement of electrons. This fact is important to the operation of a pn junction and practically all semiconductor devices.

In the area near the junction where the limited combining of the majority carriers takes place, the resistivity is high. In fact, this section acts much like a dielectric insulating material between two plates of a capacitor. In the case of the pn junction, the line of the majority carrier in each segment acts as a capacitor plate, as shown in Fig. 1-15.

Much happens when external bias is applied between a piece of n-type semiconductor positioned back to back with a piece of

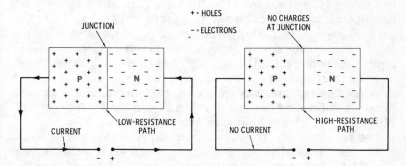

Fig. 1-16. The pn junction.

p-type material, as shown in Fig. 1-16. The motion of charges depends on the polarity of any external voltage applied across the junction. When a negative voltage is applied to the n-type material and a positive voltage to the p-type material, there is a motion of charges (current). The negative potential on the n-type material repels the negative charges, driving the electrons toward the junction between the two segments. In a similar manner, the positive potential on the p-type material drives the positive particles toward the junction. Consequently, there is a free motion of charges across the junction and a low-resistance conducting path results. The junction has been forward biased.

The activity is quite different when the p-type material is made negative with respect to the n-type material. In this case the negative charges are drawn toward the positive terminal. In a similar manner the positive charges are drawn toward the negative terminal. Therefore, charges are pulled away from the junction, and there is no motion of charges between the two segments. Thus, a continuous charge motion (current) is not established because of the very high resistance of the junction under this bias condition. In this case the junction is said to be reverse biased or back biased.

The pn junction just described has a low resistance when it is forward biased, permitting a high current. When it is reverse biased, it has a high resistance, and little or no current results. External current can be in only one direction and, therefore, a pn junction can function as a diode detector or rectifier.

When a pn junction is back biased, the majority carriers are kept back from the junction. However, there are minority carriers available, and to them the junction appears as though it is forward biased. Although this reverse resistance of the junction is high compared with the forward-bias resistance, there is a minority charge flow (Fig. 1-17B). This current is low. At low operating temperature it is usually considered insignificant. However, at high operating temperature it is a factor that must be considered. Even when forward biased, a rise in temperature increases the number of carriers, and the external current increases correspondingly.

When the reverse bias exceeds a certain value, a complete breakup of the electron bonding occurs and the reverse current rises sharply. This is known as avalanche current.

The response of a typical pn junction is shown in Fig. 1-18. After the forward bias voltage is made to exceed the barrier potential, the forward current begins its rise. The higher the forward bias voltage, the higher the junction current (up to

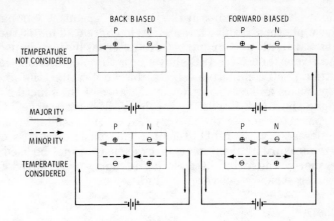

Fig. 1-17. Minority carrier flow.

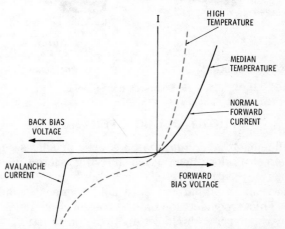

Fig. 1-18. Voltage/current characteristics of a pn junction.

normal operating limits). Current is low for the back-bias condition and changes very little with an increase in the reverse voltage up to the avalanche breakdown value. At this potential there is an abrupt increase in junction current.

The influence of temperature rise is shown by the dashed curve. Note that for a given increase in forward bias, there is a higher junction current. In the back-bias case, there is a current increase too. In effect, the efficiency of the pn junction as a unidirectional current device is poorer as a result of heat.

A silicon pn junction can be made to operate as a photovoltaic cell or as a photoconductive diode (Fig. 1-19). In Fig. 1-19A the pn junction acting as a photovoltaic cell responds to the impinging light photons arriving at the junction. The energy imparted by these photons results in an electric current. This is the arrangement used in solar panels. No external battery is used.

The circuit of Fig. 1-19B shows how the pn junction can be used as a photoconductive cell. In this application, the photon energy biases and controls the conductance of the junction. The stronger the light intensity, the lower the resistance of the path connected across the battery. A current results. This current increases with light intensity. This basic circuit is often used to measure light intensity. However, it cannot be used as a solar energy converter because more energy is removed from the battery than is delivered by light energy.

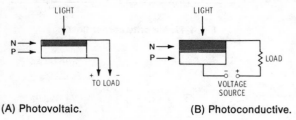

(A) Photovoltaic. (B) Photoconductive.

Fig. 1-19. Solar-cell connections.

PHOTOVOLTAIC OPERATION

The influence of arriving photon energy produces a minority current effect as does heat. Each arriving photon, in the case of silicon, frees a valence electron. Enough energy is imparted to the electron to transform it to a high-energy conductance electron. This free electron and its associated holes are attracted by the barrier shield set up at the pn junction. The electron and hole travel in opposite directions from the majority carriers and actually set the direction of the photovoltaic current present in the output circuit (Fig. 1-20). When the external load is connected, the charge flow in the output circuit is from the p-type material to the n-type material. Electron motion is directly proportional to the light intensity.

The response of a typical silicon photovoltaic cell to radiant energy is shown in Fig. 1-21. Note the maximum response in the 8000–9000 angstrom spectrum. This particular response means that the silicon cell is an efficient converter of solar energy because the intensity of solar energy peaks at about 7000 angstroms and is at a high level between 6 and 9000 angstroms. An approximate rounded curve of solar energy intensity is shown in Fig. 1-22.

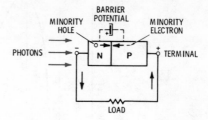

Fig. 1-20. Operation of photovoltaic cell.

Silicon cells are constructed of wafers of very high purity, one part impurity per billion. These wafers are about 0.05 centimeter thick. Such crystalline material is doped with a tiny amount of boron. Since the boron valence ring is shy an electron, the material assumes a positive charge (p-type). A diffusion process is then used to form a thin layer of n-type material. The diffusion of a small amount of arsenic with its extra valence electron establishes the negative charge of the n-type

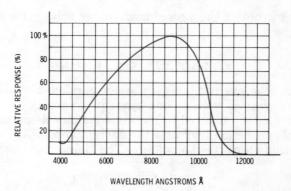

Fig. 1-21. Photovoltaic response of typical silicon cell to wavelengths of solar energy.

layer. In a practical silicon cell, the light photons pass through the thin n-type layer to the junction. Negative and positive charges are formed, with negative charges diffusing across the junction into the n-type silicon and positive charges to the p-type. These charges build up at the connecting leads and flow in the external circuit when a load is connected.

Typical output characteristics of a silicon photovoltaic cell are shown in Fig. 1-23. Note that the top curve corresponds to the standard STC value of 100 mW/cm² solar intensity. The dashed vertical line indicates the optimum float-charge voltage provided by the cell. This value is somewhat higher than the full-charge voltage of the battery. Light cells are connected

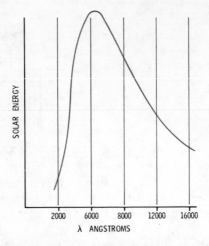

Fig. 1-22. Solar energy level as a function of wavelength.

in series to obtain the desired float-charge voltage for a particular battery.

A PRACTICAL SOLAR PANEL

Basic cells are combined in series-parallel combinations to obtain a desired voltage and current capability (Fig. 1-24). They are interconnected in the same manner as batteries. The parallel connection increases the current capability; the series connection steps up the voltage. The Spectrolab model 12-3.0

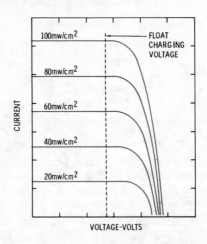

Fig. 1-23. Response of silicon photovoltaic cells.

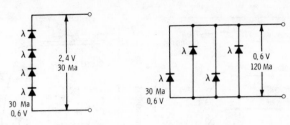

Fig. 1-24. Typical outputs of cells connected in series and parallel.

(Fig. 1-25) consists of ten modules to obtain a float-charge voltage of 13.5 volts and a current capability of 3 amperes. Each module consists of 120 individual cells in a series-parallel combination that produces 13.5 volts at 300 milliamperes (three parallel groups of 40 in series). As shown, each module is encapsulated in a strong, flat, plastic tube. The individual cells are often suspended in a clear, inert silicone compound and hermetically sealed. Mounting is very important as the panel is exposed in all types of weather.

Courtesy Spectrolab

Fig. 1-25. Solar panel.

This light-energy converter has been designed for charging 12-volt lead-acid and nickel-cadmium storage batteries. Performance characteristics of the 12-volt 6-ampere model are given in Fig. 1-26. Note that the top curve is again the STC

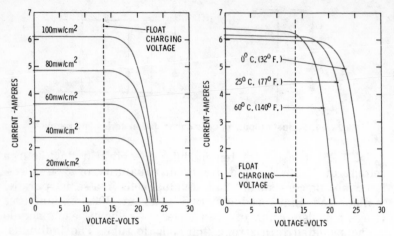

Fig. 1-26. Response of Spectrolab 12-V, 6-A array.

standard. The float charge is 13.5 volts. Under standard sunlight it delivers a bit more than 6 amperes.

The second set of curves shows the performance of the array at standard intensity for various temperatures. Note that at the float-charge voltage, there is very little change in output current for a substantial change in ambient temperature.

Table 1-1 gives specifics on various practical models, giving size, current capability, and weight. Additionally, the number of ampere-hours generated per day based on solar conditions is of assistance in deciding what combination is most suitable for a particular location. Note for the 12-volt, 6-ampere model that 26 ampere-hours is made available daily under average solar conditions.

Fig. 1-27. A two-inch-diameter silicon solar cell.

Courtesy International Rectifier Corp.

Table 1-1. Spectrolab 12-V Array Data

STANDARD 12-VOLT[1]LEC™ ARRAYS

Array Model No.	Approximate Dimensions (inches)	Current Output (Amps)[1]	Amp-Hours Generated Under Various Per Day Solar Conditions[2]			Approx. Weight (lbs.)
			Poor	Average	Good	
12V .3A	37 x 3 x 3	.3A	1.1	1.3	1.6	3.8
12V .6A	37 x 6 x 3	.6A	2.2	2.6	3.1	9.3
12V .9A	37 x 9 x 3	.9A	3.3	3.9	4.7	14.0
12V 1.2A	37 x 12 x 3	1.2A	4.4	5.2	6.3	18.6
12V 1.5A	37 x 15 x 3	1.5A	5.5	6.5	7.9	23.3
12V 1.8A	37 x 18 x 3	1.8A	6.6	7.8	9.4	28.0
12V 2.1A	37 x 21 x 3	2.1A	7.7	9.0	11.0	32.6
12V 2.4A	37 x 24 x 3	2.4A	8.9	10.4	12.6	37.3
12V 2.7A	37 x 27 x 3	2.7A	10.0	11.7	14.1	41.9
12V 3.0A	37 x 30 x 3	3.0A	11.1	13.0	15.7	46.6
12V 3.6A	37 x 37 x 3	3.6A	13.3	15.6	18.9	55.9
12V 4.2A	37 x 43 x 3	4.2A	15.5	18.2	22.0	65.2
12V 4.8A	37 x 49 x 3	4.8A	17.7	20.8	25.2	74.5
12V 5.4A	37 x 55 x 3	5.4A	20.0	23.4	28.3	83.8
12V 6.0A	37 x 61 x 3	6.0A	22.2	26.0	31.5	93.1
12V 6.6A	37 x 67 x 3	6.6A	24.4	28.6	34.7	102.4
12V 7.2A	37 x 73 x 3	7.2A	26.6	31.2	37.8	111.7

1. Minimum current output under Standard Test Conditions (STC)
   Intensity = 100 mw/cm²; Temperature = 0°C. to +60°C.
2. Usable energy generated for use in a solar power supply system
   with lead acid storage batteries under typical conditions and
   based on annual mean solar radiation data for various locations
   in the contiguous United States.

Silicon solar cells are available in a variety of shapes and sizes. Some have active areas as small as 0.01 inch square; others go up to 2-inch-diameter sizes (Fig. 1-27). Various other shapes and arrangements are shown in Fig. 1-28. Included is a panel made from a series-parallel grouping of round cells. A rugged case provides protection from the mounting-site environment.

Comsat and Solarex have developed the so-called violet photovoltaic cell. Efficiencies in excess of 15% have been obtained. This high-efficiency cell, shown in Fig. 1-29, has a higher sensitivity than the normal cell, extending into the short-wavelength spectrum (violet region). A number of factors contribute to the rise in efficiency.

Improved efficiency results from three major improvements—a shallower junction, an improved collector arrange-

Fig. 1-28. Various solar cells and solar panels.

ment for interfacing between cell and contact, and the use of a more transparent protective coating. The very thin diffusion layer more readily permits photon penetration over the solar energy frequency spectrum and more efficient conversion to charge motion regardless of wavelength. Such a thin diffusion layer has a higher lateral resistance. Consequently, the interface must be of a finer geometry. A pattern of fine wires serves as an effective grid because of its lower series resistance. An

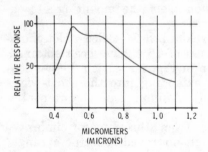

Fig. 1-29. Response of silicon violet cell.

improved coating was needed to permit easy penetration (minimum reflection) to the active cell surface. This improved transparency was gained with the use of an antireflection film that becomes an integral part of the cell structure.

CADMIUM SULFIDE CELLS AND ARRAYS

Presently the silicon photovoltaic cell is the most common light-energy converter. However, there are other photovoltaic-effect devices. In all of these, three processes are involved. Initially, positive and negative carriers must be freed by the arriving light energy. There must be a junction where excess charges can be separated, either a pn junction or a metal semiconductor junction. The freed carriers must be mobile and must remain in their separated state for a time interval longer than that needed to travel to the charge-separating junction. The silicon semiconductor pn junction performs these services adequately as discussed in the previous section. The second most popular device today is the cadmium sulfide cell.

The silicon photovoltaic cell is known as a homojunction type because the n-side and p-side, except for the impurity add-on, are both made of the same material, silicon. The cadmium sulfide cell is known as a heterojunction because one side is cadmium sulfide and the other copper sulfide.

Of course, there are many combinations of both types. Some have been tried and a few have been successful. Others are without promise and some are as yet untried. There has been little work with the metal-semiconductor junction; a successful combination may be inexpensive to produce. One does not know what a really serious development effort will bring forth.

The cadmium-sulfide cell has several attractive attributes. It can be manufactured in a thin-film procedure, and there are indications that such a process lends itself to mass production. In general, the efficiency and lifetime of such a cell are less than that of a silicon cell. However, recent changes in the manufacturing process have made substantial improvement.

The center for development in the United States is the University of Delaware. Under the direction of Dr. Böer, a solar home, shown in Fig. 1-30, has been constructed. The thin-film construction of the cadmium sulfide arrangement permits its use as both a heat collector and a generator of electricity. This dual function capability is a powerful advantage. More details about Solar I are given in the last chapter.

The thin-film construction permits an active area comparable to the diameter of a human hair. The sandwichlike construction plan is shown in Fig. 1-31. First, there is a metal substrate, into which thin-film n-type cadmium sulfide is evaporated. Next, a very thin p-type layer of copper sulfide is deposited on top of the cadmium sulfide by using an ion-exchange reaction. To this is cemented a light-transference grid electrode. Sealing is by way of a Mylar sheet.

Fig. 1-30. Solar One at the University of Delaware.

Light photons pass through the transparent grid and thin p-type layer to the junction. As in the silicon type, carrier bonds are broken and minority carriers flow through the junction. A voltage develops across the two segments. This open-circuit voltage is approximately 0.37 volt. At a conversion efficiency of 7%, a current of 19 milliamperes/cm^2 is present in a short-circuited load connected across the output leads.

Again, as in the case of silicon cells, a higher output voltage is obtained by connecting many cells in series. Higher current capability is obtained by paralleling cadmium sulfide cells. The

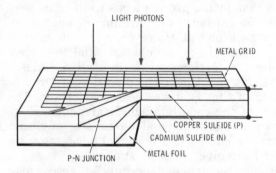

Fig. 1-31. Cadmium-sulphide (CdS) photovoltaic cell.

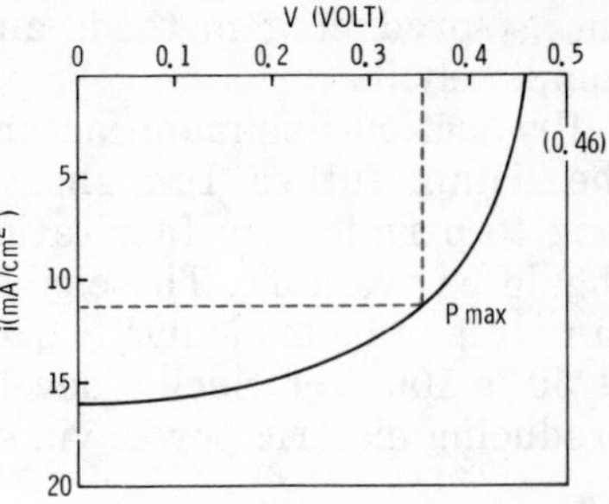

Fig. 1-32. Response of CdS cell.

chart of Fig. 1-32 shows the voltage-current relationship for a cell with a 4.3% conversion efficiency.

Tyco Solar Energy Corp. has developed a silicon-ribbon solar cell. This advance stems from their skill at producing sapphire filaments for use in sodium vapor lamps. Other firms have been licensed to use the Tyco process in the manufacture of sapphire substrates for semiconductor devices. In the manufacturing process, a molten mass of material is shaped into a single crystal by a capillary-action die which conducts the metal to its top surface, determining the shape of the solidifying crystal through surface tension force. There are many cost advantages to this technique as compared with the conventional crystal-growing process. The growth process is referred to as edge-defined, film-fed growth, or EFG.

For application in solar cells, crystal ribbons 18 inches long, 1 inch wide, and 0.008 inch thick have been produced. The optimum thickness for a silicon-ribbon solar cell is 0.004 inch. No difficulty in producing this thickness is anticipated.

More recently, 6-foot lengths have been made with a solar cell efficiency of 10%. This indicates that very long lengths are indeed feasible. The comments of a Tyco official are, "While there is clearly much work to be done to develop the technique, the cost estimates indicate that even at a very modest level of production of the order of 20 megawatts/year, cost per peak kilowatt can be expected which is very attractive compared to conventional power producing methods. The investment required to develop the technique to the point where these costs can be achieved is very modest compared to that needed to develop, say, nuclear fusion or other advanced concepts. The reduction to manufacture of the Tyco technology for solar cells, achievable within 7–10 years at a cost of $50 to $100 million, will provide an important source of fuel-free, pollution-free electricity which will greatly reduce the burden on conventional energy-producing methods and may eventually displace them completely."

Production-line manufacture of thin-film solar cells is not in the distant future. In a short time, ribbon lengths sufficiently long to permit easy fabrication into conventional solar panels should be available. These ribbons (only ten mils in thickness) can already be made five to six feet long and it is hoped as long as 50 to 100 feet. Such a development might reduce the cost of producing electric power via solar cells by a factor of 100.

SELENIUM CELLS

The selenium cell was one of the first photovoltaic devices. It appeared prior to the tremendous breakthrough of the silicon pn junction. It was used widely in instrumentation and in various types of light-operated devices. Efficiency is substantially lower than that obtained with silicon and cadmium sulfide cells.

The selenium cell begins with a steel or aluminum base plate. To this is bonded a thin layer of selenium. The cell is approximately 0.03 inch in thickness. The metal serves as the positive electrode, the active surface of the selenium cell being negative. A metallic ohmic contact is usually located somewhere along the perimeter of the cell and serves as a link between the external circuit and the selenium.

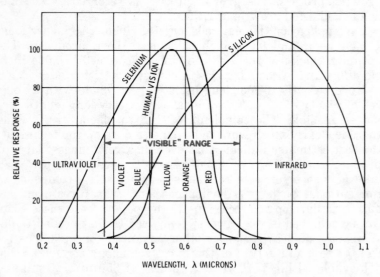

Fig. 1-33. Spectral response of photocells.

The metal base makes the cell rugged. The cell can be shaped into many configurations and sizes. This makes it especially attractive for light-measuring devices and light-sensitive industrial applications. A special attraction of the selenium cell is its maximum sensitivity in the visual frequency range (Fig. 1-33). Note that its response peak is in the orange spectrum of the visual light range. Silicon peaks in the infrared spectrum.

There is much to be learned and much experimental work to be done in the development of solar cells. Gallium-arsenide cells with 18% efficiency have been processed. Gallium is reasonably abundant. Schottky-barrier devices show promise and consist of two thin semiconductor layers and an ultra-thin metallic coating. It is a new energy ball game.

CHAPTER **2**

Wind Energy and Conversion

Winds precede and follow the cyclic weather fronts that move across the continent from west to east. In most places there are more hours of wind than calm. On an annual basis, the wind velocity in southeastern Pennsylvania is greater than 8 miles per hour for more than 60% of the year as compared with times of light wind and calm. The winds blow across your backyard, sometimes whistling and howling at high velocity. The conversion of this energy to electricity can serve many useful needs.

The general plan of a wind-energy to electrical-energy conversion system was shown in Fig. 1-3. Systems must be planned according to specific needs in a way that is quite different from the usual electrical procedures. Energy must be stored for use during those time intervals of calm and low wind velocity. You must know about the weather conditions in your area. You must know what quantity of electricity (ampere-hours and/or kilowatt-hours) you require. This is a clue as to how much energy must be stored in batteries.

The conversion of wind energy to mechanical rotation and then to active or stored electricity has every indication of becoming a complex science. Windmills have been available for generations but have only been conceived and used in rather simple forms. Types of blades and rotors respond in different manners to wind. There are many arrangements now under investigation. How can the windmill be used to best advantage when it is operated as a wind generator? Can systems be designed to deliver a constant regulated output despite substantial changes in wind velocity? How can wind-generators be intermixed expediently with other forms of solar energy?

45

TAMING THE WAYWARD WIND

Wind characteristics and velocities in the area of the mounting site are important data in planning any sort of a wind generating system. Your nearest weather station can supply you with useful information. Similar data can be obtained from the National Climatic Center, Federal Building, Asheville, North Carolina 28801.

The *Climatic Atlas of the United States* is available from the U.S. Government Printing Office, Washington, D.C. 20402, for $4.25. You can also subscribe to the *Climatological Monthly Data* for your state or any group of states with which you are concerned. Subscription information can be obtained from the National Climatic Center.

Table 2-1, a chart of annual percentage frequency of wind speed groups and the mean speed, is taken from the *Climatic Atlas of the United States*. This information will give you some idea of wind velocities in your area if you select the station nearest to your site. These are average figures and may not apply exactly to your particular site. Exact figures for the intended mounting site can only be obtained by using an accurate wind-velocity meter operated near the mounting site and at the same height as the wind generator. In general, higher velocities are obtained at higher altitudes. Therefore, if your altitude is higher than that of the local weather station, you can expect a somewhat higher wind average. Of course, mountaintop sites are ideal. Although there is a law of diminishing returns, relative to costs, higher wind velocities are reached with an increase in wind-generator tower height.

Of great importance is the positioning of the wind-generator relative to nearby obstacles. The mounting position should be kept clear of and above trees, buildings, and other obstacles. If you cannot go above certain obstacles, try to maintain at least a 500-foot clearance. Try to obtain a minimum 15-foot clearance above nearby obstacles.

From Table 2-1 note that the mean wind speed for Philadelphia, Pennsylvania is 9.6 mph. This is a reasonable figure for the southeast corner of the state, with its rather flat and slowly decreasing altitude toward the near sea-level altitude of the Delaware River as it passes Philadelphia. Mean speeds are somewhat higher at locations that are north and west of the city.

Note from the chart that for 62% of the year, wind velocity exceeds 8 mph and for 27% of the year, it exceeds 13 mph. These are attractive figures for wind-generator operation. It

Table 2-1. Winds by Speed Groups and Mean Speed
(Annual Percentages)

State and Station	0-3 mph	4-7 mph	8-12 mph	13-18 mph	19-24 mph	25-31 mph	32-38 mph	39-46 mph	47 mph and over	Mean speed mph
ALA. Birmingham	27	22	30	17	3	1	*	*	*	7.9
Mobile	7	28	38	20	6	1	*	*	*	10.0
Montgomery	31	29	27	12	2	*	*			6.9
ALASKA, Anchorage	28	35	25	11	2	*	*	*		6.8
Cold Bay	4	9	18	27	21	14	5	2	*	17.4
Fairbanks	40	35	19	5	1	*	*	*		5.2
King Salmon	11	20	30	24	10	4	1	*	*	11.4
ARIZ. Phoenix	38	36	20	5	1	*	*	*		5.4
Tucson	18	35	30	14	3	1	*	*		8.1
ARK. Little Rock	12	30	39	16	2	*	*	*		8.7
CALIF. Bakersfield	35	30	24	10	1	*	*			5.8
Burbank	52	26	18	4	1	*	*	*		4.5
Fresno	30	41	22	7	1	*	*			6.1
Los Angeles	28	33	27	11	1	*	*			6.8
Oakland	26	28	28	16	2	1	*	*	*	7.5
Sacramento	15	28	31	18	5	1	*	*	*	9.3
San Diego	28	38	28	6	*	*	*			6.3
San Francisco	16	21	26	22	11	3	*	*	*	10.6
COLO. Colorado Springs	9	27	38	19	6	2	*	*	*	10.0
Denver	11	27	34	22	5	2	*	*	*	10.0
CONN. Hartford	13	26	32	24	6	1	*	*	*	9.8
D.C. Washington	11	26	35	22	5	1	*	*	*	9.7
DEL. Wilmington	15	31	30	19	4	1	*	*	*	8.8
FLA. Jacksonville	10	33	35	18	3	*	*	*		8.9
Miami	14	30	34	20	2	*	*	*	*	8.8
Orlando	18	28	32	17	4	*	*	*		8.6
Tallahassee	33	36	23	7	*	*				6.1
Tampa	9	31	40	16	2	*	*	*	*	8.8
West Palm Beach	9	22	36	27	6	1	*	*		10.5
GA. Atlanta	13	24	36	21	6	1	*	*		9.7
Augusta	36	29	25	9	1	*	*			6.3
Macon	10	26	46	16	2	*	*	*	*	8.9
Savannah	12	34	37	14	3	*	*	*	*	8.4
HAWAII. Hilo	7	34	43	15	2	*	*			8.7
Honolulu	9	17	27	32	12	2	*	*	*	12.1
IDAHO, Boise	15	30	32	18	4	1	*	*		8.9
ILL. Chicago (O'Hare)	8	22	33	27	8	2	*	*	*	11.2
Chicago (Midway)	7	26	36	25	5	1	*	*	*	10.2

State and Station	0-3 mph	4-7 mph	8-12 mph	13-18 mph	19-24 mph	25-31 mph	32-38 mph	39-46 mph	47 mph and over	Mean speed mph
ILL. Moline	14	23	32	24	7	2	*	*	*	10.0
Springfield	7	22	28	27	12	3	1	*	*	12.0
IND. Evansville	19	23	32	21	5	1	*	*		9.1
Fort Wayne	9	23	33	25	8	2	*	*	*	10.9
Indianapolis	9	22	34	26	7	2	*	*	*	10.8
South Bend	7	21	35	30	7	1	*	*		10.9
IOWA, Des Moines	3	17	38	29	10	3	1	*	*	12.1
Sioux City	10	20	31	25	10	4	1	*	*	11.7
KANS. Topeka	11	19	30	27	10	2	*	*	*	11.2
Wichita	4	12	30	31	16	5	1	*	*	13.7
KY. Lexington	8	25	39	22	6	1	*	*		10.1
Louisville	17	28	31	20	3	1	*	*		8.8
LA. Baton Rouge	17	29	34	17	3	*	*	*		8.3
Lake Charles	19	31	29	17	4	1	*	*	*	8.5
New Orleans	16	27	32	19	5	1	*	*	*	9.0
Shreveport	12	26	37	21	4	1	*	*	*	9.5
MAINE, Portland	10	30	33	22	4	1	*	*	*	9.6
MD. Baltimore	7	24	39	22	6	2	*	*	*	10.4
MASS. Boston	3	12	33	35	12	4	1	*	*	13.3
MICH. Detroit (City AP)	8	23	37	26	5	1	*	*	*	10.3
Flint	16	26	32	22	3	1	*	*	*	9.0
Grand Rapids	14	23	32	25	5	1	*	*	*	9.8
MINN. Duluth	6	15	33	31	11	4	1	*	*	12.6
Minneapolis	8	21	34	28	9	2	*	*	*	11.2
MISS. Jackson	33	25	26	14	2	*	*	*		7.1
MO. Kansas City	9	29	35	23	5	1	*			9.8
St. Louis	10	29	36	21	3	1	*	*	*	9.3
Springfield	4	13	34	32	13	3	1	*	*	12.9
MONT. Great Falls	7	19	24	24	15	9	3	1	*	13.9
NEBR. Omaha	12	17	29	28	11	3	*	*		11.6
NEV. Las Vegas	18	26	25	20	8	3	1	*	*	9.7
Reno	52	20	13	10	4	1	*	*	*	5.9
N. J. Newark	11	25	34	24	5	1	*	*	*	9.8
N. MEX. Albuquerque	17	36	26	13	5	2	*	*	*	8.6
N. Y. Albany	23	24	27	21	4	1	*	*		8.6
Binghamton	11	23	35	25	5	1	*	*	*	10.0
Buffalo	5	17	34	27	13	3	1	*	*	12.4
New York (Kennedy)	6	17	35	28	10	3	*	*	*	12.0

Table 2-1. Winds by Speed Groups and Mean Speed (Annual Percentages)—cont'd

State and Station	0-3 mph	4-7 mph	8-12 mph	13-18 mph	19-24 mph	25-31 mph	32-38 mph	39-46 mph	47 mph and over	Mean speed mph
N. Y. New York (La Guardia)	6	15	30	31	12	4	1	*	*	12.9
Rochester	8	22	34	25	9	2	1	*	*	11.2
Syracuse	14	27	30	23	5	1	*	*	*	9.7
N. C. Charlotte	20	32	31	14	2	*	*	*		7.9
Greensboro	20	32	31	14	2	*	*	*	*	8.0
Raleigh	18	33	34	14	2	*	*	*	*	7.7
Winston-Salem	19	22	33	21	4	1	*	*		9.0
N. DAK. Bismarck	14	20	27	24	12	3	1	*	*	11.2
Fargo	4	13	28	31	15	7	2	*	*	14.4
OHIO, Akron-Canton	7	25	35	26	5	1	*	*	*	10.4
Cincinnati	11	27	36	22	4	1	*	*	*	9.6
Cleveland	7	18	35	29	9	2	*	*	*	11.6
Columbus	26	23	29	18	4	1	*	*	*	8.2
Dayton	8	25	36	23	6	2	*	*	*	10.3
Youngstown	7	26	36	24	6	1	*	*	*	10.3
OKLA. Oklahoma City	2	11	34	34	13	6	1	*	*	14.0
Tulsa	9	24	34	26	7	1	*	*	*	10.6
OREG. Medford	47	31	14	6	2	*	*	*	*	4.6
Portland	28	27	25	16	4	1	*	*	*	7.7
Salem	25	32	28	13	2	*	*			7.1
PA. Harrisburg	28	31	25	13	3	1	*	*		7.3
Philadelphia	11	27	35	21	5	1	*	*	*	9.6
Pittsburgh	12	26	34	22	4	1	*	*		9.4
Scranton	11	33	35	18	2	*	*	*		8.8
R. I. Providence	11	20	32	28	7	2	*	*	*	10.7
S. C. Charleston	12	28	35	19	4	1	*	*		9.2
Columbia	25	35	26	12	2	*	*			7.0
S. DAK. Huron	10	18	29	29	10	3	1	*	*	11.9
Rapid City	15	22	28	21	10	4	1	*	*	11.0
TENN. Chattanooga	39	25	24	11	1	*	*			6.1
Knoxville	29	29	25	12	4	1	*	*	*	7.5
Memphis	14	26	34	20	5	1	*	*		9.4
Nashville	27	31	25	14	2	*	*	*	*	7.2
TEX. Amarillo	5	15	32	32	12	4	1	*	*	12.9
Austin	13	25	34	23	5	1	*	*		9.7
Brownsville	10	17	25	30	14	3	*	*	*	12.3
Corpus Christi	11	16	26	33	12	2	*	*		11.9
Dallas	9	21	32	28	9	1	*	*		11.0

State and Station	0-3 mph	4-7 mph	8-12 mph	13-18 mph	19-24 mph	25-31 mph	32-38 mph	39-46 mph	47 mph and over	Mean speed mph
TEX. El Paso	10	22	32	22	9	4	1	*	*	11.3
Ft. Worth	4	14	34	34	10	3	*	*	*	12.5
Galveston	4	13	39	33	10	2	1	*	*	12.5
Houston	6	18	36	28	10	2	*	*	*	11.8
Laredo	6	15	32	34	12	1	*	*		12.3
Lubbock	4	11	33	34	13	5	1	*	*	13.6
Midland	9	22	38	26	4	1	*	*		10.1
San Antonio	18	23	32	22	4	1	*	*		9.3
Waco	3	14	36	35	10	2	*	*		12.5
Wichita Falls	5	22	41	27	5	1	*	*	*	10.5
UTAH, Salt Lake City	12	33	36	14	4	1	*	*	*	8.7
VT. Burlington	24	24	28	22	2	*	*			8.3
VA. Norfolk	14	23	30	25	6	1	*	*	*	10.2
Richmond	14	37	36	11	1	*	*	*		7.8
Roanoke	31	22	23	17	5	2	*	*		8.3
WASH. Seattle-Tacoma AP	13	16	35	26	8	2	*	*	*	10.7
Spokane	17	38	27	14	3	1	*	*		8.1
W. VA. Charleston	29	37	25	8	1	*	*			6.2
WIS. Green Bay	8	22	32	26	10	2	*	*	*	11.2
Madison	15	22	30	23	7	2	*	*	*	10.1
Milwaukee	8	17	31	30	11	3	1	*	*	12.1
WYO. Casper	8	16	27	27	13	7	2	*	*	13.3
PACIFIC, Wake Island	1	6	27	48	17	2	*	*		14.6
P. R. San Juan	15	28	27	25	4	*	*	*	*	9.1

is these latter higher-velocity winds that are so effective in generating electrical power. Details on the author's wind-generator are given in Chapter 5.

Specifics are lacking regarding the expected increase in wind velocity with height. Apparently it is quite a variable quantity and subject to fluctuations. An approximate and sometimes relevant figure suggests an increase of 50 to 60% with a ten-fold increase in height.

A general equation for determining wind velocity at a specific height when the wind speed is known at a reference height is:

$$V_X = V_F \left(\frac{h}{30}\right)^{1/N}$$

where,

V_X is the velocity at desired height,

V_F is the velocity at reference height of 30 feet,

h is the height above ground,

$1/N$ is the estimated power law increase. The quantity substituted for N depends upon terrain conditions. In open farm country and level or slightly rolling terrain, a factor of 7 can be substituted for N. In level and rolling country with numerous obstructions, such as in suburban areas, the factor is 5. An estimate of 3 is appropriate for city outskirts and near suburbs and areas with large obstructions.

The intermittent nature of wind is also a factor. If there are potentially long intervals of calm during the year, they must be considered in planning the energy-storing block of the wind generating system. For example, in southeastern Pennsylvania the late summer months can have a sequence of calm, high-humidity days accompanied by only an occasional light breeze. It is significant that these calm days are also sunny ones and long ones. Therefore, any incorporated solar panel facility can provide supplementary power. In radiocommunications and other limited power applications, the wind/solar combination is an attractive one.

Maximum wind velocity, although not a factor in generating power, is significant in terms of the structural stability of the tower and wind generators.

WIND SPEED AND DIRECTION INSTRUMENTS

Wind speed and direction are important measurements in checking out a given site for possible installation of a wind-generator or for checking the performance of a wind-generator relative to wind velocity. The basic wind-measuring instrument is the anemometer. The most common type is the cup anemometer, shown in Figs. 2-1 and 2-2. An accompanying vane points into the wind, and indicates (Figs. 2-1 and 2-3) wind direction.

The modern cup anemometer is largely an electronic device. As shown in Fig. 2-1A, the rotating cups also rotate a chopper wheel. Light is transmitted to a photodiode as the slots of the chopper wheel pass below the lamp. The higher the wind speed, the faster is the rotation of the chopper wheel and the higher the average light that strikes the photodiode. Its output is in-

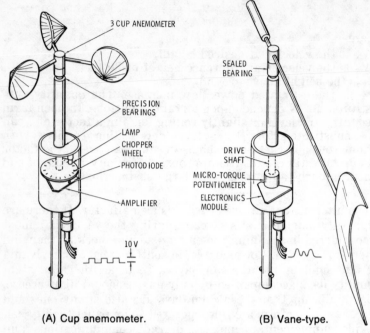

3 CUP ANEMOMETER

PRECISION BEARINGS

LAMP

CHOPPER WHEEL

PHOTODIODE

AMPLIFIER

10 V

SEALED BEARING

DRIVE SHAFT

MICRO-TORQUE POTENTIOMETER

ELECTRONICS MODULE

(A) Cup anemometer.

(B) Vane-type.

Fig. 2-1. Wind-direction sensors.

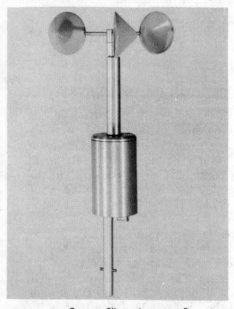

Fig. 2-2. Cup anemometer.

Courtesy Climent Instrument Co.

52

creased in level by a solid-state amplifier and is then conveyed by cable to an indicator meter or recorder.

The direction vane has a rotating shaft that turns a potentiometer. The signal released to the cable at the output of the electronic module is an indication of the potentiometer setting and the compass position in which the vane is pointing.

Two inexpensive wind speed indicators are shown in Fig. 2-4. One meter (Fig. 2-4A) is hand held at eye level with the back of the meter facing the wind. The wind pressure pushes up a white ball in the tube, indicating the wind speed in miles per hour.

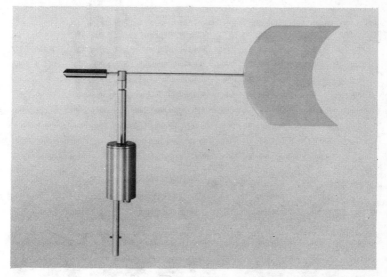

Fig. 2-3. Wind direction vane.

The other indicator (Fig. 2-4B) can be used to determine both wind direction and wind speed. In this instrument, the prevailing winds turn the weather vane and point a tube in the direction of the wind. Wind pressure in the tube pushes the red fluid, causing its level to rise and fall with changing wind speed. The indicator is also calibrated in miles per hour of wind speed. This instrument, mounted near the wind generator and with its sensor at approximately the same height, is of great assistance in checking out the performance of a wind generator.

Wind speed and wind direction instruments that keep a continuous record of these parameters are particularly important

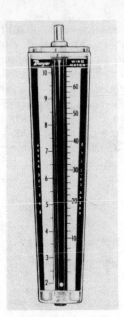

(A) Hand-held meter.

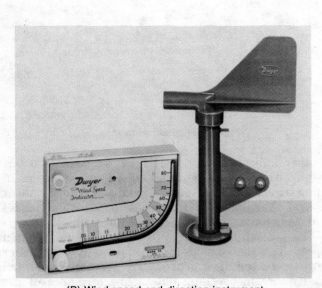

(B) Wind speed and direction instrument.

Fig. 2-4. Wind speed indicators.

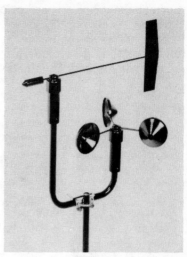

**Fig. 2-5. Wind speed and direction
sensors on same support.**

Courtesy Texas Electronics, Inc.

in checking out a site for a wind generator or in checking out
the performance of such a generator over a period of time. An
example of an anemometer cup and vane mounted on the same
support is shown in Fig. 2-5. The signals from the sensors are
transferred by cable to an indicator and recording system (Fig.
2-6). Both parameters can be read on meters, and at the same
time, their values can be recorded continuously on a chart. A

Courtesy Texas Electronics, Inc.

Fig. 2-6. Wind speed and direction indicators with recorders.

normal chart has a speed of one inch per hour and can provide a 30-day recording of the two parameters.

In large wind-generator systems, it is helpful to provide automatic switching of blade pitch, alternators, etc., as a function of the wind speed. The instrument of Fig. 2-7 is a wind speed

Courtesy Texas Electronics, Inc.

Fig. 2-7. Wind speed controller.

controller that operates off the signal from anemometer cups. A minimum and maximum wind setting can be preset to perform certain electrical switching when the wind reaches the set minimum or the set maximum.

Two modes of operation can be provided. In the automatic mode, the relay is energized at wind speeds exceeding the set point, but it automatically de-energizes when the wind speed falls below the second set point. In the latching mode, the relay energizes when the wind exceeds the set point and remains energized until manually reset. Instruments of this type will become more and more common as higher-powered and more versatile wind-generating systems are designed.

POWER LEVELS

Commercial wind generators with power levels that extend from 50 to 6000 watts are readily available. Most of the windgenerators at the upper power level of this range are manufactured overseas. Some are kept in stock by U.S. distributors. Other generators up to the 100-kilowatt range can be ordered.

Little is known about ultimate size because of our lack of experience. In terms of cost, there is certainly a crossover point where the use of two or more separate wind-generators becomes more economical than the construction of a single massive unit.

In Europe and Australia there are high-powered systems in operation. Several decades ago a 1.25-megawatt unit was operated at Grandpa's Knob in Vermont. It supplied test power and at times fed the power grid of the central Vermont Public Service Corporation between October 1941 and March 26, 1945. A fatigue failure in one of the blades shut down the machine. Materials and structural design capability have come a long way since then, and the superwind machine is now feasible. The turbine speed of the Smith-Putnam generator was 28.7 rpm, with the tip of the blade speeding at 185 miles per hour. Gears and a hydraulic coupling system linked the turbine to the generator which operated at 600 rpm. Output was 2400 volts at 60 hertz. After the 1945 shutdown, the priorities of the wartime economy forced the discontinuance of the project.

General Electric's Space Division at Valley Forge, under contract from NASA's Lewis Research Center, has begun a study of high-powered wind-generators in the 50–250 kilowatt and 500–3000 kilowatt range. The potentials of wind energy are there. A recent report of a joint National Science Foundation and a NASA panel indicated a capability of 300 billion kilowatt-hours per year from off the shore of New England, 180 billion off the mid-Atlantic, 190 billion off the Texas coast, 210 billion on the great plains, and 400 billion along the Aleutian Island chain.

A modest wind-powered electrical system consisting of several wind generators could supply electricity to small towns. Small systems capable of supplying 200 on up to 1200–1500 kilowatt-hours per month could serve homes, farms, offices, and small businesses.

Low-powered local radio stations could install their own fully active or emergency power systems using wind generators and, perhaps, solar panel augmentation. It might be that the antenna tower could also serve as the support for the wind generator. Municipal radio systems could develop electrical power in a similar manner as well as shared two-way radio and/or repeater plants.

Wind/solar electrical systems provide a decentralized means of generating electrical power. Thus, some of the horrors of massive electrical grid failure or disasters could be circumvented. There has been much talk about national self-sufficiency. We must also realize that one way of reaching toward that goal is to let local areas and individuals be self-sufficient.

Power capability versus weight is an economics factor. Researchers at Princeton University have developed a sail-wind generator using a cloth-bladed structure. One unit, weighing only 300 lbs with a 25-foot-diameter sail wing, generated between 7 and 8 kilowatts. The sail-wing blade itself is made of dacron. Tests indicate an efficiency of nearly 42%, which is climbing very close to the calculated theoretical efficiency of 59%.

The wattage rating of a typical wind generator is based on the minimum wind velocity that will produce maximum output. For example, the unit of Fig. 2-8 produces about 200 watts of output with a wind velocity of 23 mph. Wind velocities above this level produce little increase in output. In fact, automatic braking systems are included in most wind generators to hold down rotational speed and prevent breakdown from excessive wear and pressures.

SIMPLE PLAN FOR A LOW-POWERED SYSTEM

Assume that a wind-generator installation delivers approximately 48 kWh/month at 12 volts. This corresponds to 4000 ampere-hours (48,000/12) of electricity. In a practical situation this many ampere hours would be delivered to a battery-charging system (Fig. 2-9). Depending upon charge system and battery quality, something less than this number of ampere-hours would be made available for the load to be connected to the wind-powered electric system.

Let us determine if this amount of electrical energy will meet the requirements of a particular service. The system is to supply power to a radio transmitter-receiver combination. Assume that on transmit the power demand is 120 watts. At 12 volts this represents a current demand of:

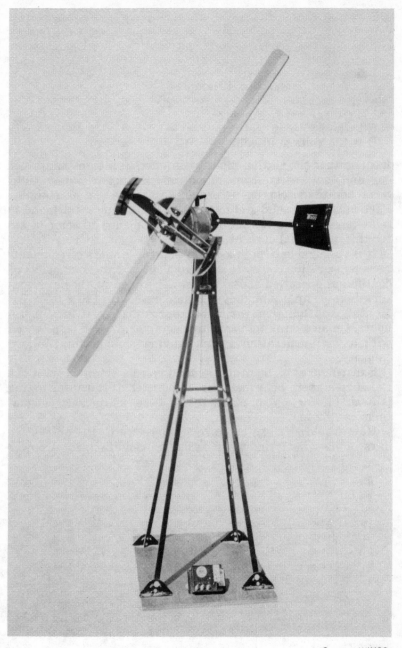

Fig. 2-8. 200-watt wind generator.

$$I = \frac{P}{E}$$

$$= \frac{120}{12}$$

$$= 10 \ \text{amperes}$$

where,

I is the current in amperes,
P is the power in watts,
E is the voltage in volts.

Assume that the station must be in operation eight hours per day and that the transmitter on-time is approximately 40%. Some additional current is drawn on receive and on standby. Particularly for solid-state equipment, this current demand is relatively low. We can assume a 50% on-time to compensate for this additional current.

It is reasonable then to assume that for 8 hours of transmit-receive operation, there is an equivalent of only four hours of continuous operation at 120 watts. Consequently, the daily consumption is approximately 40 ampere-hours ($10 \times 0.5 \times 8$). Monthly consumption would be approximately 1240 ampere-hours. On an averaging basis, this equipment could be powered with a wind-generating system having a 4000 ampere-hour capacity.

Exact figures of monthly demand can be obtained with the use of an ampere-hour meter or kWh meter. If this information is available for a year of operating time, more-exact calculations can be made.

How would this system fare if there were six consecutive days of no wind? Under this extreme condition, total power

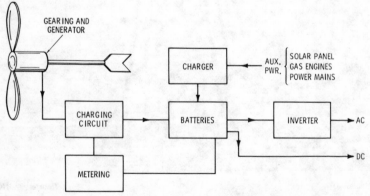

Fig. 2-9. Power source functional plan for radiocommunications.

consumption would be 240 ampere-hours (6 × 40). Your battery capacity would then have to be chosen to make this amount of electrical energy available; this would indicate a battery capacity of at least 240 ampere-hours plus an additional safety factor. A good choice might be a 360–400 ampere-hour capacity.

Again, a systematic investigation of the site area is helpful in determining the possibility of the worst conditions; that is, the longest number of consecutive no-wind days on record. The selection of ample battery capacity and the use of good-quality batteries that can withstand a rather deep discharge is important in maintaining a level voltage and good load regulation.

Your selection of the type of radiocommunication equipment is important in conserving electrical energy. Solid-state gear is ideal because no filament power is demanded. Select equipment that has very low standby current requirement and a receiver that operates at high efficiency. An example of the latter would be an audio output stage that demands current only in accordance with the strength of the amplified audio (class-B operation). Continuous-wave and single-sideband emissions are efficient modes of transmission in terms of a minimum power-supply demand. The basic equipment, associated accessories, desk lighting—all should be chosen with minimum electrical consumption in mind.

BLADES AND ROTATING DEVICES

As light as air is, it nevertheless is composed of gas molecules that have mass. A motion of these molecules is called wind. Upon striking a wind blade, sail, or other rotating device, this wind imparts energy of motion. The efficiency with which energy is transferred between wind and a rotational force depends upon the design of the wind device. For example, by setting a wind blade or propeller at optimum pitch angle, there is an efficient conversion between wind and mechanical energy.

The force of such mechanical motion or torque made available at the rotating axle depends both on wind velocity and on blade size. Available power varies as the square of the blade radius and the cube of the wind speed. For example, to quadruple power output it is necessary to double blade radius. Doubling wind speed results in an eightfold power jump.

As mentioned previously, a perfect wind generator has a theoretical efficiency of 59%. That is, 59% of the wind energy striking the area swept by its blade is converted to power. However, a figure of 50% is more likely for a well-designed windmill. Overall efficiency, considering all factors, may be in the 20–30% range.

The fundamental equation that determines the wind force, in watts, that impinges on a slim two-blade propeller is:

$$P = 0.005 \ AV^3$$

where,

A is the area in square feet covered by the blade as it rotates (equivalent to πr^2),
V is the wind velocity in miles per hour,
P is the power in watts.

Table 2-2, which follows, is quite practical and presents an approximation of the usable power that can be derived from an efficient two-blade windmill in terms of wind velocity in miles per hour and blade diameter in feet. An approximate overall efficiency of 30% has been assumed. The figures show how the available power increases with the blade diameter and the wind velocity.

Table 2-2. Output Capability in Watts for Efficient Two-Blader

Blade Diameter (Feet)	OUTPUT (Watts)			
	10 mph	15 mph	20 mph	25 mph
6	40	130	320	650
12	160	520	1280	2600
15	260	800	2000	4000
20	470	1450	3600	7200

Let us assume a blade with a 6-foot diameter and a wind velocity of 10 mph and determine the wind power striking the two-blade windmill. Calculation is as follows:

$$P = 0.005 \times 3.14 \times (3)^2 \times 10^3 = 141 \text{ watts}$$

Recall that the theoretical efficiency of the windmill is 59.3%. Let us assume a propeller-generator efficiency of 50%. Overall efficiency then approximates 30% (59.3% × 50%). Note that 30% of the previous wattage figure is about 40 W (141 × 0.3).

Actually, the above assumes a well-designed system. It is wise to consider also that there are additional losses in succeeding parts of the system as found in the charger system, batteries, distribution system, and inverter (if used).

IMPORTANT PARAMETERS AND CHARACTERISTICS

There are a number of descriptive parameters that are used in evaluating the performance of blades and propellers. They are introduced here in the discussions of two- and three-blade propellerlike structures. The simple two-blade arrangement, shown in Fig. 2-10, consists of blade, hub, and vane. The pur-

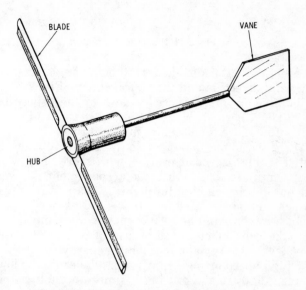

Fig. 2-10. Wind generator's basic sections.

pose of the vane is to continually turn the blade into the wind. The entire assembly must be mounted on a rotating platform arrangement. To derive maximum benefits from the wind, the deviation from true orientation should not exceed 12°.

Another important factor is the pitch angle (Fig. 2-11A). This refers to the angle of the blade relative to the wind direction. When the relative velocity is in line with the blade element, there is no transfer of wind energy to torque at the axle. For low-speed rotation of a multiblade windmill, pitch angles of 30° and higher are used. For high-speed rotation, angles are substantially smaller. Low angles are required because the airfoils of many high-speed blades stall in the 12° to 14° range. However, optimum pitch angle depends upon application, desired operating conditions, types of blade, preferred angle of attack for the airfoil, and the basic utilization of the wind-rotating system as based on the wind speed limits at the site.

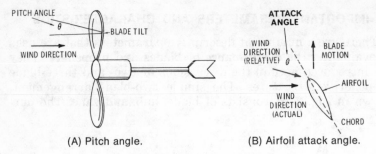

(A) Pitch angle.   (B) Airfoil attack angle.

Fig. 2-11. Blade design factors.

The angle of attack of the airfoil is the angle between the relative wind direction and the airfoil chord (Fig. 2-11B).

Windmills are often classed as low- or high-speed models. A windmill having many blades (Fig. 2-12) is usually a low-speed system of no more than several hundred revolutions per minute (rpm). A windmill having two to four propeller blades is adapted to high-speed operation with more than 300 revolutions per minute. The multiblade arrangement is a very heavy affair but develops a high torque. It is adapted for operation as a water pump and is used for other services that require a conversion between wind and some sort of mechanical activity. It does make more efficient use of a light wind in the conversion between wind energy and mechanical energy.

The high-speed or high-rpm windmill, despite its lower efficiency (especially in a light wind), is more suitable to the conversion of wind energy to electrical energy. Less weight is involved, and it presents a structure less subject to damage from strong gusty winds. Its high speed of rotation is more adaptable to low-ratio gearing of electrical generators and, in some installations, can provide direct drive of the generator axle.

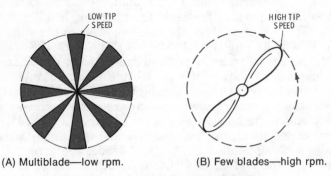

(A) Multiblade—low rpm.   (B) Few blades—high rpm.

Fig. 2-12. Low- and high-speed windmills.

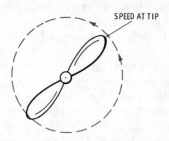

Fig. 2-13. Tip speed.

Two more factors of concern are tip-speed ratio and solidity. The tip-speed ratio (Fig. 2-13) is the ratio of wind velocity to rotational tip velocity of the mill. In fact, it is the ratio of the wind speed to the speed of motion of the very tip of the blade. A typical figure might be 0.3. Using this figure as an example, the tip speed of a particular blade when made active by a 12-mph wind would be 40 (12 ÷ 0.3) mph. For a given blade radius, a high-speed windmill has a much higher tip velocity than a lower-speed one.

Solidity is the ratio of blade area to disc area (Fig. 2-14). In the case of a two-blade propeller, it would be the ratio of the area of the circle of rotation to the area occupied by the two

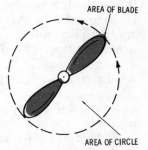

Fig. 2-14. Solidity is the ratio of blade area to disc area.

blades as they fit within the circle. A typical figure might be 0.2. Output is related to tip-speed ratio, and for a typical ratio there is an optimum solidity in obtaining the highest power output.

A blade can be flat, or it can have a twist (Fig. 2-15), just as an aircraft propeller does. A twist at the proper angle im-

Fig. 2-15. Blade with twist.

proves the power output capability of the blade. Research indicates that a 30° twist would be optimum for a tip-speed ratio of 0.3 and a solidity of 0.2.

In practice, tip-speed ratio is often stated as a whole number that compares the velocity of the blade tip with the wind speed. In this method of statement, the classic 0.3 figure would actually refer to a ratio of $3\frac{1}{3}$ (1/0.3). On this basis high-speed mills often have ratios between 5 and 8; lower-speed ones, between 1 and 3. For electrical generation, tip-speed ratios under 4 are not recommended. This limit might be made even higher as the state of the science develops. The high tip speeds bring the rotor speed closer to the required generator rpm.

Another important factor is the lift-to-drag effect. It is an indication of how well the blade is turned by the wind relative to the torque or the opposition the blade has to being set in motion by a light wind. This lift-to-drag ratio is related to the blade construction, size, and airfoil. While power output varies approximately as the cube of the wind speed, the drag varies as the square of the velocity. A high-lift airfoil does increase the power output but also raises the drag. Nonetheless, a high ratio permits a higher output at a lower wind speed.

Windmills must be braked at high velocity to prevent breakup of the rotor in a turbulent windstorm and to prevent speed-up of structural fatigue or damage to the support structure. Braking can be accomplished in a number of ways, including setting a manual brake and/or setting the blades to a stalling pitch angle. Automatic means include the mechanical brake for the rotor of Fig. 2-8 or the use of a governor-vane arrangement (Fig. 2-16). In the latter method, an excessive wind activates a spring arrangement that works in conjunction with a second vane. This second vane pulls the orientation vane to an angle that moves the platform in such a manner that the blades become end-on to the wind (again a zero pitch angle).

A third system uses an automatic method of changing the angle of the blades to a stalling pitch, much as an aircraft propeller can be feathered. This happens automatically whenever

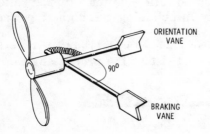

ORIENTATION
VANE

90°

BRAKING
VANE

**Fig. 2-16. Braking vane to
slow windmill.**

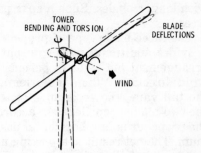

Fig. 2-17. Blade, hub, and
tower stresses.

the wind velocity exceeds a predetermined level. In fact, continuously automatic feathering arrangements have been evolved in such a manner that the rotor speed can be maintained at a constant value over a substantial range of wind velocities. This technique will be discussed further under the discussion of generators.

A windmill is placed under a variety of force stresses. These bending and twisting stresses are exerted on rotor, blades, and tower. A flapping force (Fig. 2-17) is exerted on the blade. Particularly harmful can be vibrations set up when wind forces match the natural resonances of windmill structures. Even the wake of the wind as it passes, places a stress on the support structure.

In terms of minimum cost and mechanical complexity, a direct attachment of blade to the rotor hub is attractive. This attachment, however, conveys stresses to the rotor. A more complex rotor hub is advisable for longer blades because of flapping motion and changing aerodynamic loading. Various arrangements are employed that permit hub teetering using gimbals. The cantilevered arrangement of Fig. 2-18 may become popular

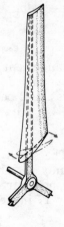

Fig. 2-18. Hingeless (cantilevered) rotor plan.

with longer blades. Such a spar provides good flexibility under stress.

The three-blade type of propeller shown in Fig. 2-19 not only provides additional power output as compared to a two-blade arrangement but reduces periods of vibration with changes in wind direction. When the orientation of the windmill follows the tail vane, the resistance to the orientation shift made by the two-blade propeller is in accordance with its position. When the propeller is in a horizontal position, this resistance is maximum. The net result is a jerking movement of the mill as it follows a wind direction change. It is true that the two-blade propeller is more economical and results in a higher tip-speed ratio. However, there are more undesirable stresses when it is compared with a three-blader. The three-blade or four-blade

Fig. 2-19. Three-blade propeller.

Courtesy Solarwind Co.

arrangement has a steady resistance as the tail vane responds to a wind direction change. Therefore, the supporting structures need be of less complex design and lower cost.

MINI-POWER WIND GENERATOR

The small multiblade wind generator of Fig. 2-20 is used mainly for charging batteries aboard small boats. It employs 14 short airfoil-section blades. It has an overall diameter of only 17 inches, and the total unit weighs only 9 pounds. This

Fig. 2-20. Aerocharge multiblade generator.

small unit is made available by Solar Energy Co. and provides an average charging rate of 250 mA at 12 volts. Charging begins at a wind speed of 10 mph and reaches a maximum output with a speed of 25 mph (Fig. 2-21). The windmill rotator

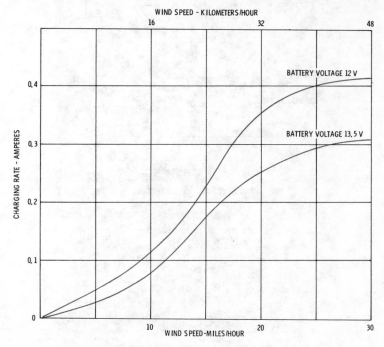

Fig. 2-21. Aerocharge generator performance chart.

drives the alternator directly without the use of gears. Rotation at average charging rate is about 300 rpm. The orientation assembly is sealed and consists of two sealed bearings.

There is an associated control box which houses the rectifier that converts the generator output to dc current. The control box also includes an indicator light which glows when the unit is charging.

Installation is simple for all types of boats. The unit can also be used on land for low-powered remote radio stations and can be left unattended for long periods of time. Recommended installation for land applications is at a height of 15 to 20 feet in the clear.

CHALK ROTATOR

The concept of the chalk rotator shown in Fig. 2-22 is unique, effective, and lightweight. It starts easily in a light breeze.

Fig. 2-22. Chalk spoked-wheel generator.

Early measurements indicate that it is capable of reaching an efficiency in the 50% region (recall that the theoretical maximum is 59.3%).

The chalk rotor consists of a spoked wire wheel. The structure supports light-weight sheet-aluminum blades shaped in an appropriate airfoil section. The spoked-wheel construction provides great strength despite the low weight; for example, a 15-foot-diameter wheel weighs about 70 pounds.

An important advantage of the construction is that it simplifies gearing to a generator. As an option, it is possible to extract power at the rim. Since the rim speed of the wheel (comparable to the tip speed of a conventional blade) is high, it may be used to direct-drive a generator or to drive a very simple gearing system. In fact, it is conceivable that the field poles themselves can be made a part of the rim assembly.

Table 2-3 shows the power output for spoked-wheel wind turbines at various wind speeds. Note that for a small wheel less than 8 feet in diameter, 77 watts are generated at a 10-mph wind velocity. A thirty-footer in an area with an average wind speed of 10 mph could provide all or most of the power needs of a small to modest dwelling.

PRINCETON-SAIL WIND GENERATOR

The adaptation of the Princeton sail wing to windmill research is a natural. This advanced sail wing developed by Princeton University was conceived initially for boat application and eventually for use as an aircraft wing (Fig. 2-23). Its

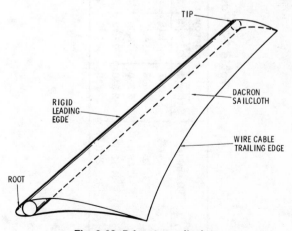

Fig. 2-23. Princeton sail wing.

Table 2-3. Power Output of Spoked-Wheel Wind Turbines

Diameter	Horsepower			Kilowatts (70%)			kWh per Month		
	10 mph	20 mph	30 mph	10 mph	20 mph	30 mph	10 mph	20 mph	30 mph
7.64 ft	.147	1.18	3.98	.077	.615	2.08	55	443	1494
15.28 ft	.589	4.71	15.91	.307	2.46	8.31	221	1772	5979
30.56 ft	2.360	18.86	63.64	1.230	9.84	33.22	886	7087	23919
61.12 ft	9.340	75.40	254.57	4.920	39.37	132.89	3543	28348	95678
122.23 ft	37.710	301.70	1018.00	19.683	157.46	531.00	14172	113375	382649

structure is simple, lightweight, and efficient. Materials are inexpensive and permit a more simplified support structure than conventional blades. A sail wing has a rigid leading edge. The root section is attached to the hub of the rotor. Additionally, both the tip and the root are connected by a trailing-edge wire cable. This is fastened to a wrapped-around sail. The sail is cut in such a manner that its trailing edge shape is set by the tension of the trailing-edge cable. A taut wing with a simple structure (Fig. 2-24) is the result. However, the wing deforms and responds to a load in accordance with the velocity and angle of the wind, developing an effective aerodynamic characteristic.

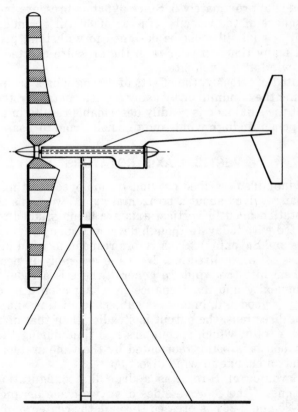

Fig. 2-24. Princeton sail-wing wind generator.

Of importance is its high lift-to-drag ratio. The sail wing has a lift coefficient and a gentle stall characteristic that compare favorably with those of a conventional hard wing and blade.

The sail wing has the same load-carrying capability and approximately one-half the structural weight of even the most basic and inexpensive conventional wing.

Furthermore, it has the high efficiency of a sophisticated hard blade despite the favorable economics of its own structure and its accompanying support tower. In fact, its weight is such that a 25-foot-diameter blade of a two-blade windmill is possible before aerodynamic effects become troublesome. For windmills larger than 25 feet, three or more blades are advisable.

A study of wind conditions on the continental United States indicates that the maximum ratio between maximum and average wind is approximately 6. Since dynamic pressure increases as the square of the velocity (factor of 36), it is understandable that a windmill must be designed to withstand pressures that are many times greater than the pressure exerted by the average wind at a given site.

As stated previously, the effects of strong winds are reduced by braking the windmill or by using a pitch control system. The fact that the sail blade is readily deformable results in a twisting component in high wind, and this holds rpm to a safe value.

VERTICAL-AXIS ROTATORS

The windmills described previously employed rotating structures that revolved about a horizontal axis. Two plans that involve rotation about a vertical axis are being studied and experimented with today (although developed many years ago by Darrieus and Savonius). Such rotors respond to wind pressure regardless of wind direction. No vane assembly is needed to orient them into the wind. In general, there is a reduction in both complexity and maintenance by using such a structure. Efficiency is good and, in an area subject to gusting and changing wind directions, the output is steadier than that of a horizontal-axis rotor, which encounters loss time during those intervals when it is being reoriented by the vane system to accommodate a change in wind direction.

The Savonius or S-rotor is a drumlike configuration. Air striking one of the concave sides of a two-blade arrangement (Figs. 2-25 and 2-26) is pressed through the center vent of the rotor to the back of the convex side. This activity sets up the rotational pattern shown in Fig. 2-26. It has been a successful wind rotor and has been used widely for ship propulsion, ventilators of many types, and water pumps. One design has been used as a successful ocean-current meter.

There is much to be learned about the Savonius S-rotor. What is the most efficient and/or effective aspect ratio (ratio of height to diameter)? How do the shape, number of blades, and venting system affect operation?

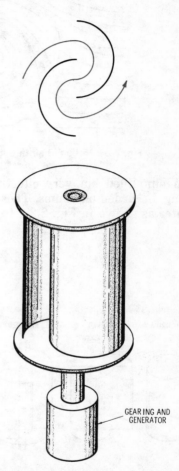

Fig. 2-25. Savonius S-rotor.

GEARING AND
GENERATOR

There is every indication that the Savonius has a high starting torque. This means that in general application it will begin to rotate and generate energy at a low wind speed. The speed of rotation is slower, but more power can be made available. The slow speed does necessitate a higher gear ratio if the Savonius is to be used as a wind electric generator. However, the mechanical structure can be simplified because the generator can be mounted at the base by using a long vertical axis.

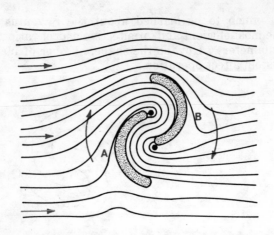

Fig. 2-26. Motion of air through vent of Savonius S-rotor.

A simplified but very effective S-rotor can be constructed using discarded oil drums. These are halved and positioned off center as shown in Fig. 2-27. In its simplest form, the drum is

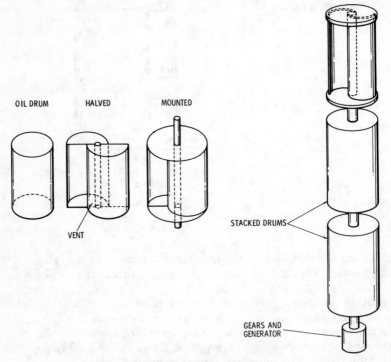

OIL DRUM   HALVED   MOUNTED

VENT

STACKED DRUMS

GEARS AND
GENERATOR

Fig. 2-27. Oil drum S-rotor.

not reshaped and the venting system remains as is. However, higher starting torque and efficiency may be obtained by making the necessary changes to obtain a more ideal vent. Multivane venting systems and the effects of using more than two vanes are essentially unexplored.

Considerable experimentation is being done with the Darrieus or catenary vertical-axis rotor shown in Fig. 2-28. Tests with both two- and three-blade types are underway. In this construction, flexible airfoil blades are used. Under action of centrifugal and aerodynamic forces, the blades assume a catenary configuration. (There is a bulging at the equator and a

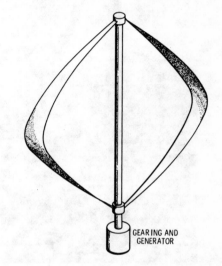

Fig. 2-28. Catenary rotor.

GEARING AND
GENERATOR

flattening at the poles.) Extensive bracing is not needed, and the supporting structure can be quite simple. Bearings at the top and bottom (Fig. 2-29) and a base-mounted gear and generator system complete the basic structure. A guy-wire system provides additional support for the entire structure. In one experimental unit, a 15-foot-diameter blade generated 1 kilowatt of power from a wind speed of 15 mph. At a wind speed of 30 mph, the output jumped to 8 kilowatts. As an assist in the manufacturing process, the blades were made of consecutive straight sections as shown. Fastened together, a reasonably catenary shape can be synthesized.

Another vertical-axis rotor is shown in Fig. 2-30. This is a very simple structure, with the wind surface being placed at the end of horizontal support rods. Either a cup or trough con-

Courtesy National Research Council, Canada
Fig. 2-29. Three-blade catenary.

figuration can be used. The cup has been used for many years, and it is the basic configuration of the wind anemometer.

The catenary vertical-axis rotor is not self-starting. A starting vane or other arrangement starts the initial rotation of the catenary blades.

The vertical-axis wind generator developed by Sandia Laboratory combines the Savonius and Darrieus concepts to obtain self-starting with a catenary vertical. This is accomplished by placing two 3-blade Savonius cups at the top and bottom of the power-generating catenary section (Figs. 2-31 and 2-32). The two Savonius buckets provide the high starting torque needed to start the rotating system in a light wind.

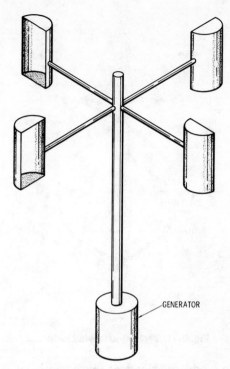

Fig. 2-30. Simple vertical-axis rotators.

GENERATORS

In a wind-generator system there is a windmill rotor that must be coupled mechanically to an electric generator (see Fig. 2-9). This can be an ac generator alone or an ac generator with appropriate mechanical or electrical means for converting ac electricity to dc electricity. Usually, but not always, a gear arrangement is needed to match the high rpm of the generator to the relatively slower rpm of the windmill rotor.

In most systems the rpm of the windmill rotor varies with the wind velocity over a substantial range. A gear ratio must be selected to accommodate wind velocity, windmill rotor rpm, and generator rpm for the most efficient and effective operation at the wind-generator site. Automatic blade pitch control and other means can be employed in more-complicated windmills to minimize the extent of the rpm change with wind velocity.

The choice of generator size is very much a function of the power generated and the highest usable wind speed. On one hand, the generator must not have too low a power rating to

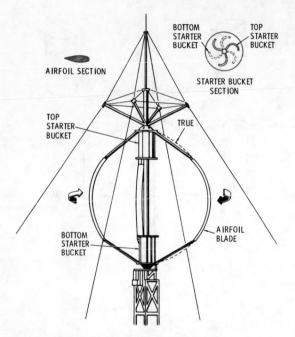

Fig. 2-31. Vertical-axis wind turbine.

prevent heating and overload at high wind velocity. On the other hand, the power rating should not be too high in terms of the useful wind capability of the site to reduce weight and economize on tower construction. There must be an optimum choice in obtaining an efficient total energy output.

Various generator drive methods are depicted in Fig. 2-33. In the direct-drive method, windmill rotor and generator rotor have the same rpm. An old and no longer manufactured Jacobs' wind generator used this method. It was a 110-volt generator with a 2000-watt rating, capable of supplying 400–500 kilowatt hours per month. Discounting the power consumed by the hot-water heater, the author's dwelling does not demand this level of power per month. The windmill and generator rotor varied from 125 to 225 rpm with wind speed levels of 7–20 mph. A 6-pole shunt generator was employed.

Meshed gear arrangements are common (Fig. 2-33B). Typically a windmill with a rotational speed of 300 rpm is stepped up to a generator rpm falling between 1500 and 3600.

Regular and V-belt drives are practical. Pulley wheel diameters are set for the proper step-up ratio. It is feasible to operate more than one generator by using suitable pulley and belt

Fig. 2-32. Vertical-axis wind generator.

arrangements. This technique is particularly suitable to the vertical-axis type of rotator.

Various types of generators have been used and are used in wind-generator systems. The two-pole ac generator is practical. Mechanical or electronic means can be used to make the conversion between ac and dc electricity when desired. Multipole generators as well as the newer constant-rpm systems are practical. There is every indication that the alternator will take over, especially in handling low powers up to the 4000–8000 watt range. For low-power applications in the hundreds of watts, the auto alternator is effective and low cost. One can anticipate the development of a variety of alternators especially fitted for wind-generator application.

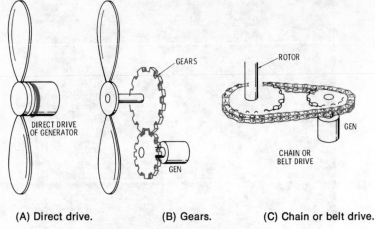

(A) Direct drive.   (B) Gears.   (C) Chain or belt drive.

Fig. 2-33. Methods of driving generators.

GENERATOR TYPES

Actually the direct-current generator using carbon brushes is common for low-powered units. Brushes and commutator provide a dc output voltage. The old-time Jacobs' system, which is quite practical for today (discussed in Chapter 5), employed large brushes with an anticipated lifetime of 15 years or more.

The modern alternator is less expensive to make because of its wide application in automobiles. Permanent magnets provide the required magnetic field in the pm-type of alternator. A stationary winding with a diode provides the field current for the brushless types. Included diodes make the conversion between ac and dc.

In wind-generator applications, the propeller speed is low and in the 100–300 rpm range. A low-speed generator to go with such a system is large, heavy, and expensive. Its lifetime is good because of its low rotational speed. Most low-powered systems employ some sort of gearing arrangement as mentioned previously. Thus, the generator speed is faster than the propeller speed, and good output can be obtained with wind speeds as low as 15–18 mph. Some sort of blade-speed control system (governor) limits blade speeds to the 25–30 mph range.

Virden Perma-Bilt Company sells a modified generator that provides 110–120 volt, 60-Hz ac output (Fig. 2-34). These are designed principally for application with a V-belt drive and pulley arrangement linked with the alternator, fan, or crankshaft of an auto. However, these two-pole generators have ap-

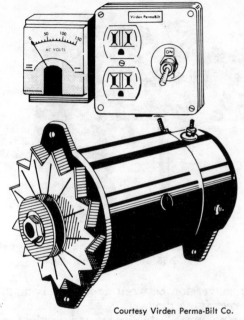

Courtesy Virden Perma-Bilt Co.

Fig. 2-34. Modified alternator for 110-V ac output.

plications in wind-generator systems. A voltage regulator could be obtained and used to maintain a reasonably constant 120-volt output, although alternator frequency may deviate from 60 hertz as a function of armature speed.

The permanent magnet self-excited type, as well as the more common types that employ dc field current, can be obtained from Virden. In the case of the latter, it is easier to exert control over the generator output by using manual or automatic control of the field current. Wattage ratings of the units extend from 1500 watts to 4000 watts.

A Quirk 2000-watt generator for 110-volt dc electrical systems includes an automatic controllable-pitch propeller. The blade arms are mounted on sealed ball bearings, the blades feathering automatically in the wind. Shown in Fig. 2-35 are the propeller hub, geared generator, and rotating platform.

BASIC AC GENERATOR

The electrical power distributed for public use is ac power. Some wind generators generate ac power; others generate ac power, which is converted immediately to dc power. In the lat-

83

Courtesy Quirk of Australia
Fig. 2-35. 2000-watt generator.

ter case, the conversion between ac and dc is made with commutators or rectifying diodes.

Alternating-current electricity is produced by the interaction between magnetic fields and electric conductors. There can be no electromagnetic induction unless there is a change in magnetic flux or physical motion. Voltages induced from such changes do not have a constant amplitude. Practical generators induce a changing voltage in the conductors.

In its simplest form, an alternating current can be produced by simply rotating one or more turns of wire in a magnetic field (Fig. 2-36). This arrangement demonstrates the principle involved in the generation of ac electricity. The flux density is

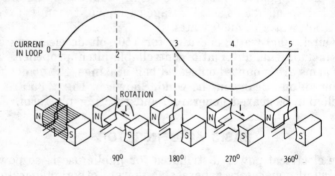

Fig. 2-36. Basic ac generator—two pole.

a constant that depends on the strength of the permanent magnet or electromagnet. The length of the conductor in the magnetic field is also constant. The loop is revolved at a constant speed by some external turning mechanical force, such as the blade and hub of a wind generator.

A factor that does change as the coil is rotated is the angle at which the conductor moves in relation to the magnetic lines of force. The angle changes through 360° as the coil is rotated. In the first position (shown in Fig. 2-36) corresponding to zero degrees, the rotating conductors are shown moving parallel to the lines of magnetic force. During a brief period, the conductors are not cutting any lines of force, so no current is induced. The voltage represented by the sine wave at zero degrees is zero. As the dark and light sides of the conductor continue to move, they begin to cut magnetic lines of force. The dark side moves down past the north pole, and the light conductor moves up past the south pole. Voltages of opposite polarity are induced in the sides of the loops. An external load placed in the loop would have an electron flow from the negative dark conductor to the positive light conductor.

At the 90° position, the maximum voltage is induced because magnetic lines are now being cut at the fastest rate. The conductors are moving at right angles to the magnetic lines, or the angle between conductor motion and the field magnetic lines is 90°. Because of the rapid flux change for this brief instant, there is maximum current present in an external load connected across the loop ends. The resulting sine wave shows that a maximum voltage is developed at the 90° position and that electrons are leaving the dark wire and entering the light one.

The conductors continue to rotate, moving toward the 180° position where the conductor motion will again be parallel to the magnetic lines. The induced voltage decreases, and the external current once again falls to zero. As the rotation continues past 180°, there is a change in the polarity of the induced voltage because the dark side now moves up past the south pole and the light side moves down past the north pole. The direction of the current present in the external load changes as well. Induced voltage and current of opposite direction continue to increase as more and more lines are cut. A second maximum is reached at 270°. Once again the conductors are cutting the field lines at right angles and therefore at the fastest rate. However, the peak induced voltage is now of opposite polarity. The induced voltage and electrical current decrease as the rotation continues, falling away to zero when the dark side of the coil reaches the top and the light side reaches the bottom. This final quarter-turn of rotation returns us to the original position, and a 360° rotation has been completed. This represents one cycle of alternating current.

So long as the loop of wire is rotated, cycle after cycle of induced voltage is produced. Alternating-current electricity is being generated by electromagnetic induction. The number of ac waves generated in a second is called the frequency. When an ac generator is attached to a windmill, the actual frequency of the ac voltage generated varies with the wind speed. Thus, in a high-speed wind the frequency of the generated waves would be greater than in a lesser wind.

In more-complex wind-generating systems, there is automatic regulation of the speed of rotation and a constant-frequency ac voltage can be generated. Such a complex system is normally not employed for modest electrical requirements. Rather it is far simpler to convert the ac voltage to dc voltage, which can then be used to charge batteries or deliver energy to other storage systems.

The simple generator of Fig. 2-36 is called a two-pole generator because it has a north and a south pole associated with the magnetic field. A simplified drawing of a four-pole ac generator is shown in Fig. 2-37. It differs because there are pairs of north and south poles associated with the magnetic field. In this arrangement, two sine waves are generated with each complete rotation of the rotor. In the case of two-pole, four-pole, or multiple-pole generators, there must be a means of removing the electrical power from the rotating rotor conductors. This is accomplished with the use of so-called slip rings that rotate with the rotor. Two electrical conductors called brushes press against the rotating slip rings and serve as a means of removing the electrical energy (Fig. 2-37).

How can a simple ac generator be used to deliver a dc output? This is accomplished by modifying the slip ring into a two-segment or split ring, the two sections being insulated from each other and the shaft. This forms the very simple commutator of Fig. 2-38. In operation the split-ring arrangement reverses the armature coil connections to the external circuit at the very same instant that the direction of the generated voltage in the coil reverses. Each time the coil is in a vertical position the output brushes exchange split rings. Consequently, the current direction in the load remains the same although the current in the coil has switched directions.

It should be noted that the amplitude of the output voltage does vary, and a so-called pulsating dc voltage is developed in the output. In a simple generator such as this, a following filter can be used to smooth out the pulsation and develop a dc output voltage. In a practical generator, however, there are more coils and more commutator bars which keep the load current more constant in amplitude, smoothing out the ripple variations.

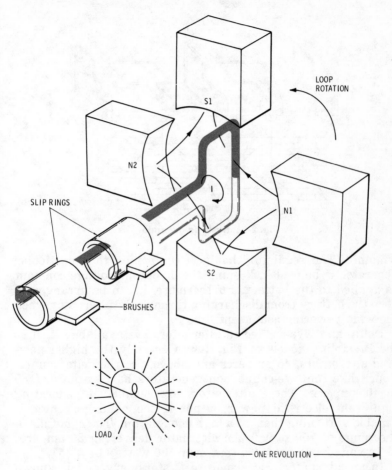

Fig. 2-37. Basic four-pole generator.

ALTERNATORS

The auto alternator along with its regulator finds application in wind-generator systems. Suitable alternators can be purchased from auto supply houses or searched for in junk yards. A pulley is already attached to the alternator, and it can be belt driven with a suitable pulley connected to the rotor of the

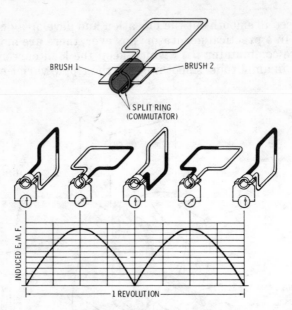

Fig. 2-38. Basic dc generator.

windmill. The regulator that is a companion to a particular generator is helpful in keeping the dc output voltage constant as applied to the battery. Furthermore, it can be arranged to keep the battery from discharging through the alternator when the wind power is insufficient.

Delta- and wye-connected alternators are available, such as the Prestolite models of Fig. 2-39. Generally the higher-powered alternators, 45 amperes and above, use the delta connection, while the wye-connection generators have ratings of 30 to 40 amperes. These alternators are capable of generating significant electrical power output if the wind generator is capable of turning them at a high enough rpm. The rpm falls in the range of 750 to 950. An alternator delivering 35 amperes at 14 volts has a power output near 500 watts (14×35).

In an alternator, the armature is stationary and is known as the stator; the field becomes the rotating member, known as the rotor. The field current is low and energizes the rotor through rotating slip rings and small brushes. The stator is fixed, and the high dc output can be conducted to the load circuit through direct leads and connections. No rotating commutator is required.

Basically the three-phase stator winding would give ac outputs 120-degrees related. However, as shown in Fig. 2-39 there

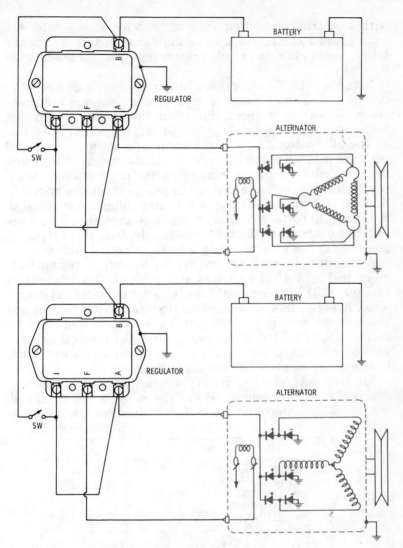

Fig. 2-39. Prestolite delta- and wye-connected alternators.

are full-wave rectifying diodes associated with each of the windings. As a result, a dc output is obtained.

The wiring of the alternator and regulator is the same as that used in automobiles. In Fig. 2-39 the switch replaces the usual ignition switch and indicator lamp. The switch in a simple wind-generating system can be operated manually. If automatic switching is desired, it can be used in conjunction with a control circuit that responds to wind speed. After the wind speed has risen to a value high enough to generate the torque needed to drive an alternator, the switch is closed automatically.

Note that the F connection of the regulator is connected to the field winding of the alternator through appropriate rings. When the switch is closed, this connection makes certain that field current is supplied from the battery to permit operation of the alternator. In fact, alternator operation is controlled with the field current because it forms the magnetic field that interacts with the stator winding as the rotor is spun by the wind-generator belt attached to the pulley. When the system is built up to normal operation, a constant voltage is supplied to the charging battery. The alternator also then supplies an appropriate current to the field winding. In fact, this current is made to vary so as to maintain a constant output voltage.

A detailed drawing of the Prestolite voltage regulator is given in Fig. 2-40. Two relays are included. One is a circuit breaker (CB) that removes the battery from the control system when it is not in operation. This prevents battery drain and protects the alternator from any battery connected with incorrect polarity. The voltage regulator (VR) keeps the charge voltage constant under normal operating conditions and under a changing load placed across the battery.

The contacts of the circuit breaker form a single-pole, double-throw arrangement. When the relay is not energized, the moveable contact and upper contact are closed, and the bat-

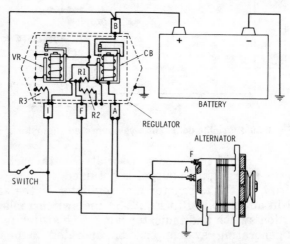

Fig. 2-40. Alternator charging circuit showing regulator schematic.

tery is not connected to the alternator. When the switch is closed, a path from the battery is connected to a tap on the circuit-breaker winding. Current of opposite direction is present in the winding, and the relay contacts remain in the same position. However, current is supplied to the field winding of the alternator. The output from the alternator builds up and is supplied to the A terminal of the circuit breaker. This opposes the current in one portion of the circuit-breaker winding. Finally, the current level is such that the circuit breaker pulls down. In so doing, contact is made between the moveable contact and the lower fixed contact, connecting the alternator and battery together. This cycle of events cannot happen if the battery is connected with incorrect polarity.

The voltage regulator also has single-pole, double-throw contacts. It employs a vibrating reed-type moveable contact. The winding of the VR relay connects between ground and the circuit-breaker yoke to the right of the regulating resistor, R1. Resistors R2 and R3 are protective and prevent surges and improve regulator stability.

In the nonoperating position, the moveable contact is spring-held against the upper contact. If the system voltage is low, the current through the coil is not adequate to overcome the tension. However, as mentioned previously, energy is being supplied to the field coil because the switch is closed.

When the system voltage builds up, the pull against the spring is such that the contact vibrates between no contact at all (float) and the upper contact. However, the rotor field circuit completed through regulator resistor R1 supplies the necessary current to the field winding when current demand is high. When the alternator speed is high (as driven by the wind generator) and when the load is low, the moveable contact is pulled down against the lower stationary one. In this vibrating mode, the end of the field winding is grounded on a momentary basis and output drops. Thus, the regulator maintains a constant charge voltage despite a changing load on the battery and, in the case of a wind-generating system, despite reasonable changes in the alternator rpm with wind speed variation.

A Motorola alternator with an electronic regulator is shown in Figs. 2-41 and 2-42. Again, "delta-connected" and "wye-connected" generators are shown along with the full-wave diode rectifiers. Also, the conventional rotor winding and slip-ring arrangement are included. The alternator includes an isolation diode. This diode provides polarity protection and keeps the battery from discharging into the charging circuit (regulator and alternator) without the need for a relay. The

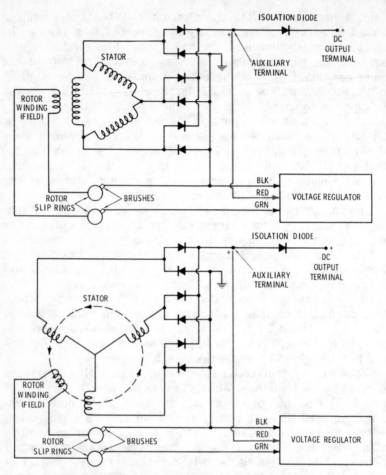

Fig. 2-41. Motorola alternator and electronic regulator.

regulator again provides a constant output voltage by controlling the current flow in the rotating field coil.

The regulator is all solid state. Its operating condition is set by the voltage that appears at the auxiliary terminal of the alternator. (The reference voltage is set by the zener diode in Fig. 2-42.) When the voltage reaches 14.4 volts, the bias at the base of transistor Q1 turns it on. In turn, transistor Q2 is switched off. This momentarily disconnects the field current and slows down the alternator. The switching action is very rapid, but a lower average field current does result.

When the auxiliary voltage falls, transistor Q1 switches off and transistor Q2 switches on. A higher average field current is supplied to the alternator. The switching of the solid-state regulator maintains a field current commensurate with the load demand and alternator rpm. Resistor R7 is a thermistor that provides temperature compensation, setting the proper operating point for the zener diode.

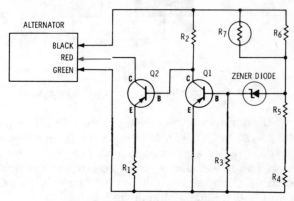

Fig. 2-42. Transistor regulator.

ADVANCED WIND-GENERATOR SYSTEMS

Wind speed is a variable quantity, and in simple wind-generating systems, output power is very much a function of the energy contained in the impinging wind. Advanced systems try to integrate this energy and spread it out more uniformly. In the case of a dc system, attempts are made to keep the dc output voltage and current capability constant. In the ac system, there is the additional parameter of frequency, and attempts are made to keep voltage, current capability, and frequency constant.

A simplified block diagram of a system under test at the University of Wisconsin is given in Fig. 2-43. A variable-frequency alternator is driven by the wind power. Its output component is rectified with a complex bridge rectifier that operates as a constant-current control. Output is supplied to a battery pump-storage facility, a dc load, and a high-capacity constant-current bridge inverter that develops a 60-hertz output.

In this plan, leverage is provided by the pumped storage. There is a feedback link between the inverter output and the pump-storage. When there is abundant power, the inverter

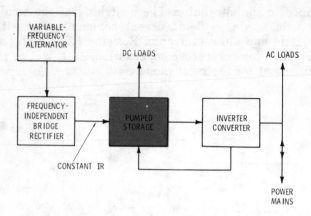

Fig. 2-43. Experimental wind-generator power system.

feeds into the pump-storage arrangement. The pump-storage is a battery system that can accept power from the wind generator or the inverter and can deliver power to the inverter or the dc load. The power distribution activity of the pump-storage depends upon power being delivered by the wind-generator system and the demands of the dc and ac loads.

Oklahoma State University has experimented with a generator having a variable-speed input and a constant-frequency output. The system employs a solid-state alternator along with a method of modulating the outputs. By using a three-phase synchronous alternator and combining the three outputs in parallel after they have been full-wave rectified, an output is produced that can make available a sine wave of a specific difference frequency (for example, 60 hertz). The output of the generator (Fig. 2-44) consists of a dc component, a component with a ripple frequency of $6F_r$ (F_r is the basic frequency of the alternator), and a full-wave rectified sine wave at the difference frequency, F_d (say, 60 Hz). The field of the generator is

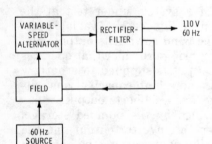

Fig. 2-44. Constant-frequency system.

excited by the ac power of the same frequency, F_d. It is this component that determines the frequency of the output.

A 10-kilowatt, 220-volt prototype system has 16 poles. Its speed of operation is 7000 rpm with a frequency, F_r, of approximately 930 hertz. The 60-Hz difference frequency at the output of the rectifier assembly is matched to the 60-Hz field to maintain a constant output frequency, regardless of loading and variable-speed input. Although designed for a rotation of 7000 rpm, it is capable of operating between 1200 and 10,000 rpm (approximately a 5:1 speed ratio). A gearing system must be employed to step-up the hub rotational speed to the range of operation of the generator.

Another approach to attaining a constant-frequency output for ac wind generators is to maintain a constant rotational speed. This can be accomplished by exact balancing of the blades in terms of pitch angle and angle of tilt. This balancing adjusts in relation to the specific backward tilt that is a function of the centrifugal force exerted on the blade in rotation. This movement is balanced by a spring force set for a predetermined rotational speed. In this arrangement, changes in wind speed or load result in a shift of the pitch angle by an amount that keeps the rotational speed constant.

SUMMARY

The science of wind blades and generators has begun anew after a lapse of 50 years. Blade types and wind systems can be selected according to wind characteristics at the site, and according to need and economic means. Blade designs themselves have come within 5 to 8 percent of the theoretical limit of 59.3%. The major thrust will be toward the economical development of the backup system, generator, and support structures—in other words, toward the establishment of a strong, safe, but light weight and economical, system.

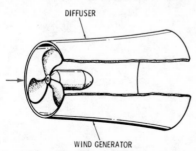

DIFFUSER

WIND GENERATOR

Fig. 2-45. Wind-generator diffuser.

In fact, work is now being done on what is called a diffuser. It is a shaped duct (Fig. 2-45) that guides a larger area of wind into the propeller causing a pressure rise. Flow velocity can be increased more than 20% above the free wind velocity. This is opposed to the 67% less than free-wind value for a nonducted windmill. Thus, the overall size of the windmill can be reduced for a given power output. However, it should be pointed out that a ducted rotor will not surpass a free rotor that has the same area as the area of the duct exit.

CHAPTER 3

Batteries, Inverters, and Electric Autos

The secondary battery is by far the most common method of storing the electric energy generated by solar and wind-generator power systems. There are primary and secondary battery cells. Secondary cells are rechargeable; primary cells are not. In a primary cell, the chemical action cannot be reversed and the consumed material cannot be regenerated. In a secondary cell, the chemical action is reversible and materials can be restored to their original form by proper recharging. Hence, the secondary cell is ideal for wind/solar generators because energy can be used to recharge batteries. At the same time or later, energy can be withdrawn from the batteries to meet the requirements of a specific electrical load.

A battery constructed of secondary cells can be discharged and charged a great many times. Just how many times such a battery can be recycled depends on its design, construction, use, and care.

Cells and/or batteries are connected in series to increase the voltage that is to be made available (Fig. 3-1). Cells and batteries are connected in parallel to increase the net current capability. Time is a factor in this consideration, and one must evaluate not only the current required by the load but just how long this current demand is to be made. The ampere-hour factor is very important in planning solar and wind power systems.

The electric car is becoming increasingly popular, especially for limited local and city driving. It is powered by secondary batteries. These batteries can be charged from the power mains. An additional degree of self-sufficiency is acquired if you completely or partially supply charge from a solar-panel

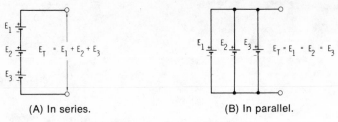

(A) In series.　　　　　　　　(B) In parallel.

Fig. 3-1. Equal-voltage batteries.

and/or wind-generator system. Even if you derive all of the charge current for your batteries from the power mains, a small passenger vehicle can be operated at a cost of 1 to 3 cents per mile, depending upon the size of the car and added electrical accessory items.

In some solar and wind installations, 110-volts ac must be made available. If your stored energy is in the form of dc battery power, it is necessary to use an inverter, which produces a 110-volt ac voltage output when supplied with a 12-volt dc input. These are the two most common input and output voltages although other dc and ac values can be employed if the inverter is designed to accommodate them.

LEAD-ACID BATTERY

The lead-acid storage battery is presently the most common energy-storing device for wind/solar power supplies. At low-power levels, the nickel-cadmium secondary battery is equally popular. The 6- and 12-volt auto storage battery is common. However, in somewhat higher-powered systems, better-grade lead-acid types are used because of their longer life and ability to be charged and discharged many times and discharged to a low value (low-valley discharge).

The common storage battery is made of lead-acid secondary cells, each making available approximately 2.1 volts (Fig. 3-2.). A 6-volt storage battery has three such cells; a 12-volt battery has six. Each cell has a removable cap permitting cell testing and the addition of water (or acid) if required.

Each cell has a number of positive and negative plates made of lead peroxide and sponge lead, respectively. The electrolyte is a solution of sulfuric acid and pure distilled water. When the battery is fully charged (Fig. 3-3), the electrolyte consists of a maximum amount of sulfuric acid and a minimum amount of water. The fully charged specific gravity of the electrolyte is nominally 1.26. This figure refers to the weight of the electro-

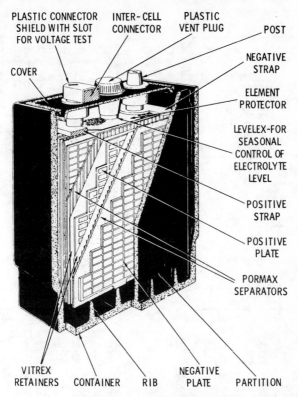

PLASTIC CONNECTOR SHIELD WITH SLOT FOR VOLTAGE TEST
INTER-CELL CONNECTOR
PLASTIC VENT PLUG
POST
COVER
NEGATIVE STRAP
ELEMENT PROTECTOR
LEVELEX-FOR SEASONAL CONTROL OF ELECTROLYTE LEVEL
POSITIVE STRAP
POSITIVE PLATE
PORMAX SEPARATORS
VITREX RETAINERS
CONTAINER
RIB
NEGATIVE PLATE
PARTITION

Fig. 3-2. Lead-acid battery.

lyte solution compared with that of an equal amount of pure water.

When power is withdrawn from the battery, the chemical action is such that the active materials of the plates combine with the acid of the electrolyte to produce lead sulfate and decrease the sponge lead and lead peroxide. When discharged, the electrolyte is composed of a minimum amount of sulfuric acid and a maximum amount of water.

To charge the battery, direct current of *opposite* direction must be passed through the cells. This opposite flow of charges reverses the chemical activity. Acid is now chemically produced at the plates and returned to the electrolyte. After a suitable charging interval, the electrolyte is restored to the original condition of a minimum amount of water and maximum of sulfuric acid.

When the battery is charged, the sulfuric acid in the electrolyte and the specific gravity are maximum. For the discharge

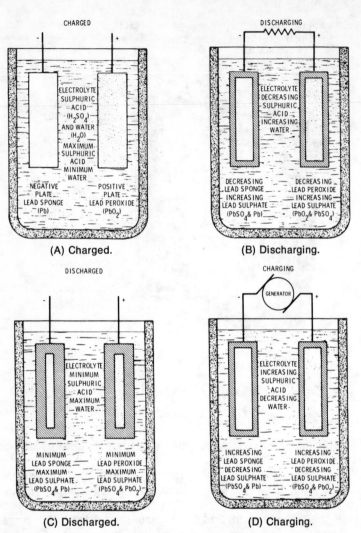

ELECTROLYTE
SULPHURIC
ACID
(H_2SO_4)
AND WATER
(H_2O)
MAXIMUM
SULPHURIC
ACID
MINIMUM
WATER

NEGATIVE
PLATE
LEAD SPONGE
(Pb)

POSITIVE
PLATE
LEAD PEROXIDE
(PbO_2)

(A) Charged.

ELECTROLYTE
DECREASING
SULPHURIC
ACID
INCREASING
WATER

DECREASING
LEAD SPONGE
INCREASING
LEAD SULPHATE
($PbSO_4$ & Pb)

DECREASING
LEAD PEROXIDE
INCREASING
LEAD SULPHATE
(PbO_2 & $PbSO_4$)

(B) Discharging.

ELECTROLYTE
MINIMUM
SULPHURIC
ACID
MAXIMUM
WATER

MINIMUM
LEAD SPONGE
MAXIMUM
LEAD SULPHATE
($PbSO_4$ & Pb)

MINIMUM
LEAD PEROXIDE
MAXIMUM
LEAD SULPHATE
($PbSO_4$ & PbO_2)

(C) Discharged.

GENERATOR

ELECTROLYTE
INCREASING
SULPHURIC
ACID
DECREASING
WATER

INCREASING
LEAD SPONGE
DECREASING
LEAD SULPHATE
($PbSO_4$ & Pb)

INCREASING
LEAD PEROXIDE
DECREASING
LEAD SULPHATE
($PbSO_4$ & PbO_2)

(D) Charging.

Fig. 3-3. Battery charge and discharge cycle.

condition, there is minimum of acid; consequently, the specific gravity is also minimum. It is apparent, then, that a hydrometer, which measures specific gravity, tells much about the operating condition of a battery. For the usual storage battery, the specific gravity at full charge is between 1.25 and 1.27 (Fig. 3-4). When the battery is discharged, the specific gravity falls to about 1.13–1.15. As the specific gravity of the electrolyte increases or decreases, the float rises or sinks. The float is calibrated for the specific gravity of the fluid.

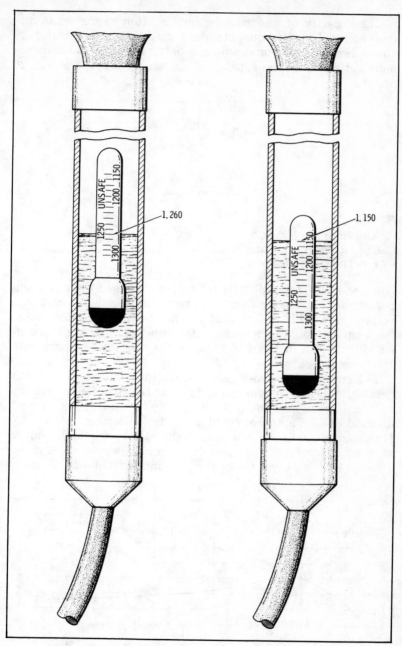

(A) Charged battery.

(B) Discharged battery.

Fig. 3-4. Hydrometer use.

The capacity of a storage battery is often expressed in ampere-hours. Usually some standard reference, 10 hours or 20 hours, is specified. For example, a battery with a 100-ampere–hour rating would be capable of delivering 5 amperes for 20 hours:

$$Ah = I \times T$$
$$T = \frac{Ah}{I}$$
$$= \frac{100}{5}$$
$$= 20 \text{ hours}$$

where,

Ah is ampere hours,
I is the current in amperes,
T is the time in hours.

Ampere-hours capability would be less or more than 100 ampere-hours for current demands greater or less than 5 amperes, respectively. This means that for an 8-ampere load demand, its capacity would be less than 100 ampere-hours, while a continuous current demand of 4 amperes would provide more than 100 ampere-hours.

The curves of Fig. 3-5 demonstrate the influence of current demand on ampere-hour capacity. For this particular battery with a 100 ampere-hour, 8-hour rating, a continuous current demand of 12.5 amperes could be made for 8 hours ($100 \div 8$). If current demand were such that the battery would discharge at a 4-hour rate, the ampere-hour rating would be only 83 (0.83×100). Current under this demand would be 20.75 ($83 \div$

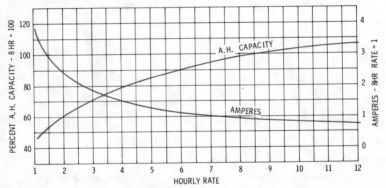

Fig. 3-5. Characteristics of 100 ampere-hour, 8-hour battery.

4) amperes. For lighter loads there would be a slight increase in ampere-hour capacity.

CONTINUOUS AND INTERMITTENT DISCHARGE

In some services there is a continuous drain on the battery, and the ampere-hours of discharge can be determined simply. One need only know what the continuous current demand is to determine the number of ampere-hours of electricity consumed. For example, with a 1-ampere current demand, the ampere-hours consumed in a day would simply be 24 ampere-hours (1×24).

In radiocommunications and in many electrical services, the current demand is a changing or intermittent one. In such applications, one must make a reasonably close estimate of ampere-hour demand over a given period of time. A precise answer can be obtained by using an ampere-hour meter. (Refer to Chapter 1.)

Let us assume that in a particular radiocommunication service, the averaging produces 15 minutes of each hour with a current demand of 5 amperes and 45 minutes with a demand of only 2 amperes. Under this manner of operation, the total ampere-hours per day would be:

$$Ah = \frac{15}{60} \times 5 \times 24 = 30 \text{ Ah/day}$$

$$Ah = \frac{45}{60} \times 2 \times 24 = 36 \text{ Ah/day}$$

$$T_{Ah} = 30 + 36 = 66 \text{ Ah/day}$$

where,

Ah is ampere-hours,
T_{Ah} is total ampere-hours.

Information of this type is valuable in determining the discharge activity of the battery system. It becomes especially important when the battery charge is the result of energy delivered by a solar panel and/or wind generator. In this situation, even the charging activity is of a varying or intermittent nature. Thus, the charging activity must be estimated according to the available light and the wind speed. A precise way to monitor both charge and discharge activities is to use two ampere-hour meters connected as shown in Fig. 3-6. Additionally, the state of the batteries should be checked, perhaps daily, by using a hydrometer.

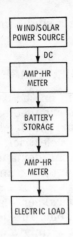

Fig. 3-6. Application of ampere-hour meter.

BATTERY CARE

A good-quality lead-acid battery in a stationary application can have a life expectancy as long as 30 years. Much depends upon the design of the battery and the relationship between cell capacity and load demand. This means the battery should not be discharged too deeply into its valley unless you employ a battery design made specifically for deep discharge. Other factors are the frequency of its cycles, the use or absence of a trickle-charge arrangement, operating temperature, efficiency of the charging system, and quality of battery maintenance.

Battery care and cleanliness are important in extending battery life. An important step in battery care is the addition of water to the electrolyte. When molecules are broken down, particularly during the charging process, they escape as hydrogen and oxygen, bubbling to the surface as gasses. This gassing causes the water loss. For this reason, there are level marks in the cell compartments. Under no circumstance should the electrolyte level go above the top mark. Also, the level should not be permitted to drop below the bottom mark. If so, the vent plug can be removed and distilled water added. The best time to make such an addition of water is when the recharge activity is about two-thirds completed.

Periodically, batteries should be dusted and cleaned. Corrosion that usually surrounds the positive terminal should be removed. Connectors should be cleaned, and vent plugs made secure. Vaseline or other battery preparations can be used to minimize terminal and connector erosion.

Ventilation is an important consideration because gassing results in the freeing of hydrogen gas which can be ignited and be explosive. Gas is released through the vent plug and occurs mainly during charge operation. Little hydrogen is released during float or trickle-charging operations. Severe gassing occurs when maximum charge voltage is impressed on fully charged batteries. Sufficient ventilation is necessary to prevent hydrogen gas from building up to a level of 3% of the air volume in the battery enclosure. Louvers in doors and/or walls, along with a small ventilating fan, eliminate the hydrogen hazard, although smoking or any source of flame or sparking should be avoided in the battery enclosure.

It is best to avoid deep cycling, although in some applications this cannot be avoided. When the battery activity is such that it produces only 80% of its rated discharge capacity, it is wise to take it out of service. This is particularly the case when it shares the load burden with other batteries. When batteries are paralleled, a bad battery can place an unnecessary load on the good battery(ies).

One of the problems of putting lead-acid batteries into operation is the handling of the sulfuric acid. The regular lead-acid storage battery is stored without an electrolyte. When it is to be pressed into service, the electrolyte—distilled water and sulfuric acid in the appropriate ratio—is added.

Modern storage batteries are available as dry-charged types. In this case, one need only add distilled water when a battery is to be added to the power system. An example is the Prestolite Oasis battery shown in Fig. 3-7. Sulfuric acid is held in special foam blocks in a concentrated state and is not activated until water is added. Thus, a long shelf life is feasible, and such batteries can be held in reserve for a lengthy period of time at the power system location.

The general plan of such a battery is given in Fig. 3-8. Water poured into a filler-opening percolates vertically through the foam block, diluting and mixing with concentrated sulfuric acid. The acid-water electrolyte rises over the plates and, when filled to the appropriate level, is then ready for immediate operation.

The small-size lead-acid batteries can also serve well in low-powered solar and wind-powered systems. Portable and other low-power radiocommunication installations for radio amateur and other services can make use of these small batteries having average dimensions of 4 inches × 5 inches × 6 inches. They are readily available at motorcycle supply houses in ratings from 4 to 20 ampere-hours. Several small-battery models have ratings as high as 30 ampere-hours. Typical systems for these small batteries are detailed in Chapter 4. They are excellent for powering solid-state equipment.

Fig. 3-7. A dry-charge secondary battery.

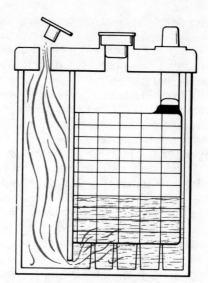

Fig. 3-8. Adding water to a dry-charged battery.

GELLED-ELECTROLYTE BATTERIES

An attractive newcomer to the lead-acid battery field is the type using a gel electrolyte (Fig. 3-9). It is truly a portable lead-acid battery that can be mounted at any angle and, in some models, even charged at any angle. Others charge more efficiently with the battery upright but can be charged at other angles with some limited decline in the total number of possible

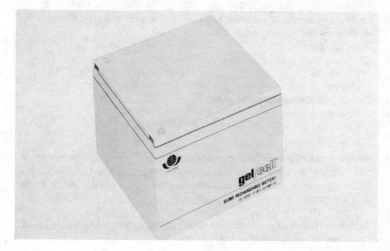

Courtesy Globe-Union Battery Co.
Fig. 3-9. A 20 ampere-hour Gel/Cell® battery.

recycles. The electrolyte is unspillable and lasts the full life of the battery, avoiding maintenance and handling problems. The battery has a one-way vent that serves as a release when there is undue gas pressure, although in this style battery there is much less gassing.

The gelled-electrolyte battery handles temperature extremes very well and is capable of performing down to $-76°F$. It is tolerant of both overcharge and deep discharge and provides long, maintenance-free shelf life. These are attractive advantages for use in solar panel and wind-generator power supplies.

Batteries may be connected in series, parallel, or series-parallel combination to obtain desired voltage and current capabilities. The Globe GC 12200, 20 ampere-hour Gel/Cell® battery has the specifications contained in Chart 3-1. Note that the rating for the battery based on a 20-hour rate is 1 ampere of current-draw for a 20-hour period, reducing the nominal battery

Chart 3-1. Specifications for Gel/Cell® GC-12200 Battery

Nominal voltage	12 volts (6 cells in series)
Nominal capacity at:	
1.0 ampere (20-hr rate) to 10.5 volts	20 Ah
1.9 amperes (10-hr rate) to 10.26 volts	19 Ah
3.5 amperes (5-hr rate) to 10.14 volts	17.5 Ah
11.0 amperes (1-hr rate) to 9.6 volts	11 Ah
Weight	16.75 pounds
Energy density (20 hr rate)	1.1 Watt-hours/cubic inch
Specific energy (20 hr rate)	14.3 Watt-hours/pound
Internal resistance of charged battery	approximately 18 milliohms
Maximum Discharge current with	
standard terminals	100 amperes
Operating temperature range:	
Discharge	$-76°F$ to $+140°F$
Charge	$-4°F$ to $+122°F$
Charge retention (shelf life) at 68°F	
1 month	97%
3 months	91%
6 months	83%

Sealed Construction: Can be operated, charged, or stored in any position without leakage of corrosive liquid or gas. Battery is protected against internal pressure build-up by self-sealing vents which pass only dry gas.

Terminal: Quick disconnect, ¼-inch. Will accept AMP, Inc. Faston "250" series receptacles or equivalent.

Case Material: High impact polystyrene, light gray color.

voltage of 12 to about 10 volts over this period. The ampere-hour rating is less when there is a greater current demand made on the battery (Fig. 3-10A). Note that a 10-ampere demand would last for only 1½ hours, the output voltage falling to about 9.8 volts. Of course, the ampere-hour rating would be somewhat greater for a current demand of less than 1 ampere.

Observe from the specifications that the battery can be left in storage six months in a charged condition with a decline to only the 82% level. The graph of Fig. 3-10B shows capacity as related to temperature and the discharge time. When discharged at the 20-hour rate, the battery provides 100% capa-

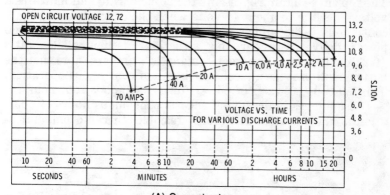

(A) Current rates.

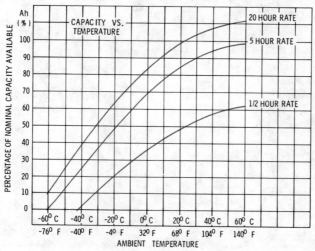

(B) Influence of temperature.

Fig. 3-10. Characteristics of 20 ampere-hour battery.

city when operated at normal room temperature (about 69°F). Capacity does fall off at lower operating temperatures as is characteristic of most batteries.

Globe makes available various size batteries that can be incorporated in cases (Fig. 3-11). These are ideal arrangements for portable radiocommunication operations. The Globe GC 1245-1 is an example and has the specifications listed in chart 3-2. This is a 4.5-ampere, 12-volt model. Details for using this type for radio-amateur QRP (low-power) operations are given in Chapter 4.

Operating characteristics are given in the chart of Fig. 3-12. The top curve again represents the 20-hour rate and indicates a voltage decline to about 10 volts when the current demand is a continuous 0.225 ampere for a period of 20 hours. The final power delivered under this situation is about 2.25 watts (0.225 × 10). If the current demand is 1 ampere, note that the battery will provide almost four hours of continuous operation. This is about 10 watts of power for a continuous four hours.

These figures indicate that several continuous hours of operation at the 5- to 10-watt level is feasible. In radiocommunica-

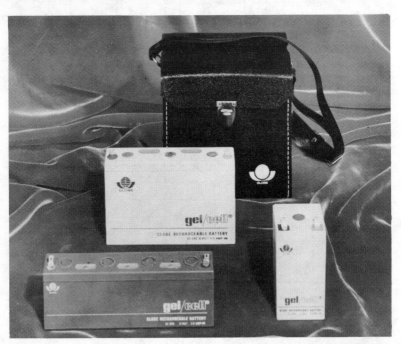

Courtesy Globe-Union Battery Co.

Fig. 3-11. Assorted-size gelled-electrolyte batteries and case.

Chart 3-2. Specifications for Gel/Cell® GC-1245 Battery

Nominal voltage	12 volts (6 cells in series)
Nominal capacity at:	
225 mA (20-hour rate) to 10.5 volts	4.5 Ah
430 mA (10-hour rate) to 10.26 volts	4.3 Ah
780 mA (5-hour rate) to 10.14 volts	3.9 Ah
2400 mA (1-hour rate) to 9.6 volts	2.4 Ah
Energy density (20-hour rate)	0.96 watt-hours/cubic inch
Specific energy (20-hour rate)	12 watt-hours/pound
Internal resistance of charged battery	approximately 60 milliohms
Maximum discharge current	80 amperes
with standard terminals	
Operating temperature range:	
Discharge	−76°F to +140°F
Charge	−4°F to +122°F
Charge Retention (shelf life) at 68°F	
1 month	97%
3 months	91%
6 months	82%

tions the bulk of the current is drawn during the transmit
mode, and the actual number of operating hours may be perhaps double, or more than double, the conservative figures
given. More information is given on this topic in Chapter 4.

The operating condition of the gelatin-electrolyte battery
can be monitored by keeping watch on the output voltage under
load. By so doing, you prevent overcharge or discharge to too
deep a level. The cases of many of these models include their
own charger.

The rated capacities (20-hour basis) for various gelled-electrolyte batteries are given in Table 3-1. Charging information

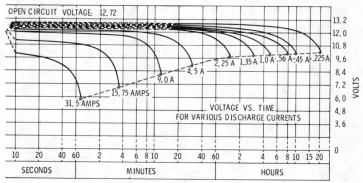

Fig. 3-12. Characteristics of 4 ampere-hour battery.

is also given. The initial maximum charge current is high when the battery is discharged. When the battery is charged, this current falls to a significantly lower value. For practically all types of rechargeable batteries, the slow charge is much preferred over the fast charge, although certain battery types are less ill-affected than others by a fast charge. In certain circumstances, you may have to sacrifice some battery life in favor of fast charges. However, in many radiocommunication applications this is not necessary, and advantage can be taken of the slow charge and, therefore, an extension in battery life can be gained.

In practical situations the battery is discharged to a specific end voltage. At this time the battery is again charged fully, then discharged, etc. In this mode of operation, the battery is on charge whenever it is not being discharged by a connected load.

There are also two arrangements used to keep a battery on continuous charge. These are known as constant-voltage charge (float-voltage charge) or a constant-current charge (trickle charge). The float-voltage system is preferred for the gelled-electrolyte batteries; the charge voltage is held constant while the current is free to vary. In contrast, the trickle-charge plan is usually employed for the conventional lead-acid storage battery. In this case there is a constant charging current while the voltage is allowed to vary.

Note in the specifications that for a 4.5-Ah, 12-volt battery, the beginning charge level is several hundred milliamperes. Full charge is indicated when the battery voltage is reading

Table 3-1. Capacity and Charging Data for Gel/Cell® Batteries

Part Number	Nominal Capacity (Ampere-Hours)	Maximum Initial Charge Current (Amperes)	Approximate Final Current (Milliamperes)
GC 210	0.9	0.15	10-20
GC 410	0.9	0.15	10-20
GC 610	0.9	0.15	10-20
GC 1215	1.5	0.25	20-40
GC 620	1.8	0.30	20-40
GC 426	2.6	0.40	30-60
GC 626	2.6	0.40	30-60
GC 1245	4.5	0.70	50-100
GC 660	6.0	0.90	60-120
GC 280	7.5	1.20	80-160
GC 680	7.5	1.20	80-160
GC 12200	20.0	4.00	100-300

14.4 volts and the charge current has dropped to a level between 50 and 100 mA. This corresponds to a final cell voltage of 2.4 volts.

When the same gelled-electrolyte batteries are to be kept on continuous charge, it is preferable that the charge voltage be held to 2.25 volts per cell, or for the 4.5-Ah version, a final voltage of 13.5 volts. Therefore, the charge must supply a constant 2.25 volts per cell (13.5 volts in the case of the 4.5-Ah, 12-volt battery).

To obtain the maximum number of recharge cycles, the on-charge voltage initially should be such that the battery charge is brought to 2.4 volts per cell. This charge should be continued until the current drops to the values shown in the tables. At this point the charge should be switched over to a float voltage of 2.25 volts per cell.

In practice, as many as 200 to 400 full charge/discharge cycles are possible. If a float-voltage charge is maintained instead of permitting complete discharge, thousands of cycles of operation are feasible.

A charger recommended for gelatin-electrolyte batteries is shown in Fig. 3-13. Remember that a constant-voltage fast or floating charge is required. A bridge rectifier supplies pulsating voltage to a series current transistor and a zener-diode regulator circuit. This circuit is recommended for charging 12-volt Gel/Cell batteries. For fast-charge activity it provides

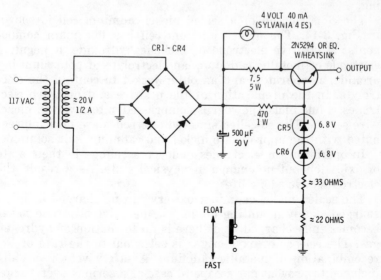

Fig. 3-13. Charger for Gel/Cell® batteries (12-volt).

exactly 14.4 volts. The constant voltage is maintained by the series power transistor and the two series-connected voltage-regulator zeners. The precise value of the constant voltage is set by the two resistors connected between the anode of the lower zener diode and common. In the float position, only one resistor is used, the output voltage being maintained at 13.8 volts. An additional resistor is inserted in the circuit when the fast-charge voltage of 14.4 volts is desired. The source of the 117-volt ac input voltage can be the power main or a solar panel and/or wind-generator power source of adequate power capability. If need be, an inverter is employed between the source and the transformer primary.

NICKEL-CADMIUM BATTERIES

The sealed nickel-cadmium battery is increasingly popular and is available in the various standard primary battery sizes and shapes although it is a rechargeable type. Small chargers are available to recharge nickel-cadmium batteries overnight. They can be recharged by suitable solar panel or wind generator systems, providing they include the necessary constant-current charging circuit. Charging technique is similar to that used for a lead-acid battery except that a higher charge voltage is usually employed and a better constant-current characteristic is needed than that required for lead-acid trickle charging.

The construction of a sealed nickel-cadmium cell is shown in Fig. 3-14. The nickel-cadmium cell has five main components: a positive electrode of nickelic hydroxide, a negative electrode of metallic cadmium, an electrolyte of potassium hydroxide, a separator, and an outer jacket to contain the electrolyte. In the preparation of the positive and negative electrodes, a fine nickel powder is compressed onto a woven nickel wire screen. The so-called sintered electrodes are now impregnated with the appropriate nickel and cadmium salt solutions.

In operation the electrodes undergo a change in their state of oxidation but no change in physical state. As a result, the electrodes have a long life.

The sealed nickel-cadmium battery can be charged and discharged at any mounting angle. The negative electrode never becomes fully charged, and there is no formation of hydrogen gas. The release of oxygen gas is set equal to the rate of the recombination of metallic cadmium. A safety vent is usually included to prevent gas escape in case of a serious overcharge. Nickel-cadmium batteries, in particular, should not be overcharged. During discharge the cadmium metal of the negative plate is oxidized to cadmium hydroxide and electrons are released to the external circuit. The nickel hydroxide reduces to a lower valence state by accepting electrons from the external circuit. These activities are reversed on charge.

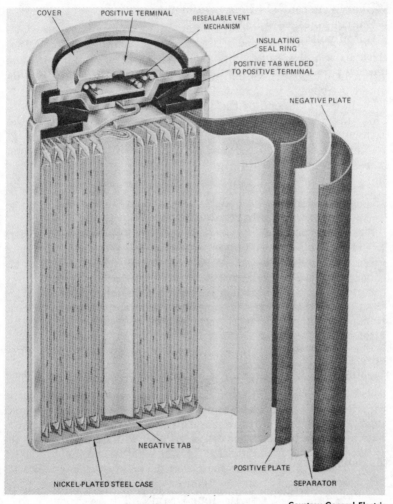

Courtesy General Electric

Fig. 3-14. General Electric nickel-cadmium sealed cell.

Sealed nickel-cadmium batteries are available with typical ratings up to approximately ten ampere-hours. Vented types are available with ratings up to 100 ampere-hours and higher.

The "D" size Eveready N56 nickel-cadmium battery, based on a 10-hour rate, has the following characteristics:

Voltage—1.25 V
Nominal capacity—2 Ah
Discharge current (10-hour)—200 mA
Charge current (14-hour)—200 mA
Charge voltage—1.35 to 1.5 V
Cutoff voltage—1.10 V

This information indicates that the battery can be operated into a load that draws 200 milliamperes for 10 hours. In this time the average voltage will have been 1.22 volts and the cutoff voltage at the end of the 10-hour period would be 1.1 volts. This states that based on the 10-hour rate the capacity of the battery is 2 ampere-hours (200 mA × 10). The curve of Fig. 3-15 can be used to determine the capacity of the battery when there is an hourly drain higher than the 10-hour rate.

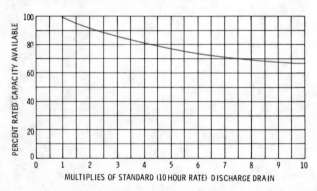

Fig. 3-15. Capacity of Eveready N56 battery at multiples of standard 10-hour rate drain.

For example, what is the capacity when the battery must deliver 400 mA to a load? Note that the abscissa is calibrated in multiples of the standard discharge drain. In the example, the multiple will be:

$$\text{Multiple} = \frac{\text{Drain}}{\text{10-Hour Drain}} = \frac{400}{200} = 2$$

From the chart it is indicated that the battery will now operate at approximately 92% of the cell capacity when the drain current is twice the standard discharge value. As a result, the available capacity in ampere-hours will be 0.92 × 2, or 1.84 ampere-hours. This means that the battery would be capable of delivering 400 mA for a period of 4.6 (1.84 ÷ 0.4) hours, at which time the battery voltage would have dropped to an endpoint value of 1.1 volts.

To obtain 12 volts at 400 milliamperes, ten of these cells would be connected in series (10 × 1.22). At the end of the discharge period, the cell voltage drops to 1.1 volts and the cutoff battery voltage would be 11 volts (1.1 × 10).

Table 3-2. Burgess Nickel-Cadmium Battery Ratings

Cell	MAH	Recommended Charging Rate for 14-16 Hrs. Charge (Milliamperes)	Trickle Charge Rate (Milliamperes)
CD1	20	2	0.2
CD2	50	5	0.5
CD8	100	10	1.0
CD3	150	15	1.5
CD4	225	22.5	2.2
CD5	450	45	4.5
CD6	450	45	4.5
CD9	900	90	9
CD10	4000	400	40
CD12	1200	120	12
CD13	2300	230	23
CD7	2300	230	23
CD14	1900	190	19

Ratings for typical Burgess batteries are given in Table 3-2 and typical General Electric batteries are given in Table 3-3. Note that charging rates are given too. Generally speaking, charging rates are usually about one-tenth of the rated discharge rate. However, the actual charging rate depends very much upon the battery design and application.

In addition to the standard types, there are fast-charge nickel-cadmium batteries and quick-charge, high-temperature,

Table 3-3. General Electric Nickel-Cadmium Battery Ratings

GE Model No.	Cell Size	Capacity @ 1-Hour Discharge Rate (Ampere-Hours)	Maximum Charge Rate (Milliamperes)
GC1	AA	0.5	50
GC2	C	1.0	100
GC3	D	1.2	100
GC4	D	4.0	350
GC5	D	1.2	250

and standby power types. Each has its own capabilities and charging procedures. Some can be charged in one-half hour. In terms of nickel-cadmium batteries, it is best to follow the procedures recommended for the particular battery type.

A basic charger is shown in Fig. 3-16. Note that the values are selected in accordance with the number of series cells that must be charged. It should be noted that the peak secondary ac voltage is significantly higher than the series battery voltage. Likewise, an adjustable potentiometer is included that permits precise control of the series charging current. This charger, which is recommended by Burgess, can be used to charge the cell types indicated in Table 3-2. Values can be adjusted for either normal charge or trickle charge.

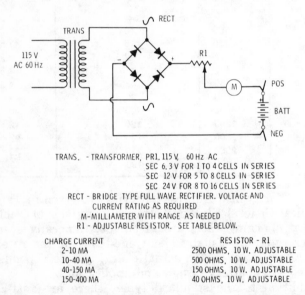

TRANS. - TRANSFORMER, PRI. 115 V, 60 Hz AC
SEC 6.3 V FOR 1 TO 4 CELLS IN SERIES
SEC 12 V FOR 5 TO 8 CELLS IN SERIES
SEC 24 V FOR 8 TO 16 CELLS IN SERIES
RECT - BRIDGE TYPE FULL WAVE RECTIFIER. VOLTAGE AND
CURRENT RATING AS REQUIRED
M-MILLIAMETER WITH RANGE AS NEEDED
R1 - ADJUSTABLE RESISTOR. SEE TABLE BELOW.

CHARGE CURRENT	RESISTOR - R1
2-10 MA	2500 OHMS, 10 W. ADJUSTABLE
10-40 MA	500 OHMS, 10 W. ADJUSTABLE
40-150 MA	150 OHMS, 10 W. ADJUSTABLE
150-400 MA	40 OHMS, 10 W. ADJUSTABLE

Fig. 3-16. Charger recommended for Burgess batteries (Table 3-2).

The cell-voltage and discharge-time characteristics of nickel-cadmium cells are interesting (Fig. 3-17). Note that the voltage on discharge holds rather constant over the discharge time until near the break in the curve. It is not advisable to let the cell voltage drop below 1.1 volts before recharge is initiated. In recharging, the cell voltage is held up and the life of the battery is extended. Thus, when nickel-cadmium batteries are used in any power source, it is important to keep watch on the voltage; discharge should be stopped immediately when the cell voltage drops close to 1.1 volt. The lowest graph shows the capacity and discharge-current characteristics of the cell. For example, it shows that with a 1.2-ampere discharge current and a cutoff of 1.1 volt, the battery has an ampere-hour capacity of about 1.12 ampere-hours.

The Eveready N86 battery consists of ten of these cells connected in series to obtain 12.5 volts (Fig. 3-18). On the basis of a 10-hour discharge at the rate of 120 mA, the capacity of the battery is 1.2 ampere-hours. Its actual discharge charac-

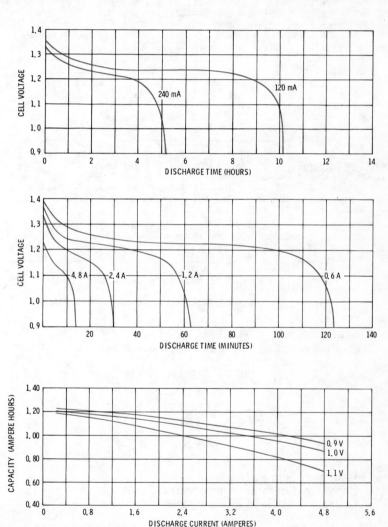

Fig. 3-17. Characteristics of a nickel-cadmium cell.

teristic indicates that the battery voltage drops to 11 volts near the 10-hour calibration line when the current demand is 120 mA.

How long will a 900-mAh cell deliver 360 mA to a load before the cell voltage drops to 1.1 volts? The 900-mAh rating

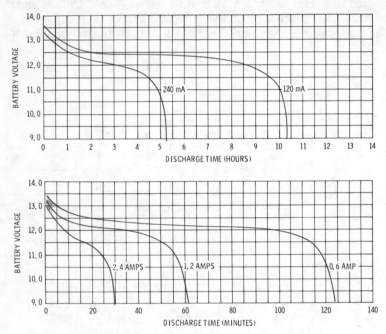

Fig. 3-18. Discharge characteristics of Eveready N86 battery.

for the CD9 cell is based on a 10-hour rate. However, with a higher current demand the hour rate is lower, and its value is:

$$\text{Hour Rate} = \frac{\text{Capacity (10-hour)}}{\text{Current Drain}}$$

$$= \frac{900}{360}$$

$$= 2.5$$

where,
Hour rate is in hours,
Capacity is in milliampere hours,
Current drain is in milliamperes.

The curve for the 2.5-hour rate (Fig. 3-19) shows that a voltage drop to 1.1 volts occurs in approximately 73% of the nominal milliampere-hour capacity. Thus, the cell will discharge to 1.1 volts in the following time:

$$\text{Service Hours} = \frac{\text{Capacity (2.5-hours)}}{\text{Current}}$$

$$= \frac{0.73 \times 900}{360} = 1.8 \text{ hours}$$

In the case of a continuous drain of 360 mA, the cell would require recharging at the end of 1.8 hours.

If this current drain of 360 mA would have to be maintained for 5 hours, what cell would you recommend? The milliampere-hour capacity for a 5-hour period would have to be:

$$\text{Capacity} = \text{Hours} \times \text{Current} = 5 \times 360 = 1800 \text{ mAh}$$

Note from Table 3-2 that two of the cells, CD7 and CD14, have capacities of 2300 and 1900, respectively. However, it is important to remember that these capacities are based on a 10-hour rate. This would rule out the CD14 almost immediately. You can determine if the CD7 would be adequate. Based on a 10-hour rate, it can supply 230 mAh for a 10-hour period. However, the current drain is to be based on a 5-hour rate, and the percentage of normal capacity from Fig. 3-19 is 90.

$$\text{Capacity} = 0.9 \times 2300 = 2070 \text{ mAh}$$

The CD7 will be adequate because its capacity based on the 5-hour rate is 2070 mAh (higher than the required 1800).

Higher-capacity nickel-cadmium batteries are of the vented type (Fig. 3-20). These must be mounted upright because they use a liquid electrolyte. Gases are generated in the battery during charge and discharge and must be leaked out through the

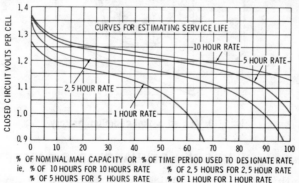

Fig. 3-19. Battery-service life curves.

vent. On occasion, an appropriate electrolyte must be added. The electrolyte is strong and must be handled carefully, more carefully than the sulfuric-acid electrolyte of a lead-acid cell—if that is possible.

Its advantages are outstanding long life on the shelf and in operation, low maintenance cost, and high resistance to shock and vibration. It holds a constant voltage over the normal discharge time.

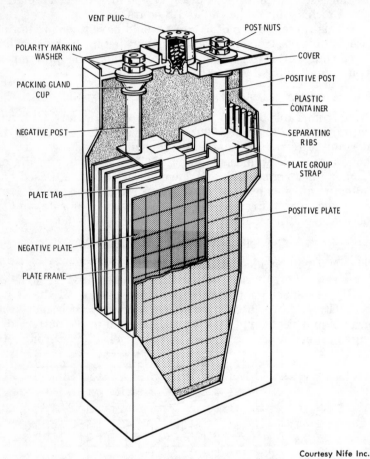

Courtesy Nife Inc.

Fig. 3-20. Vented nickel-cadmium battery.

NICKEL-HYDROGEN RECHARGEABLE BATTERY

Recently the Communications Satellite Corporation (COMSAT) developed a rechargeable battery with improved performance for use with solar panels. This battery takes advantage of the durable nickel-hydroxide electrode of a nickel-cadmium battery and the hydrogen-side electrode of a fuel cell. As a result, the capacity and life of the battery are improved over those of a comparable nickel-cadmium cell. Furthermore, the buildup of hydrogen pressure due to overcharge is avoided. A platinum catalyst is a part of the hydrogen electrode and permits rapid chemical recombination of gases. In this style of battery, a hydrogen pressure sensor can provide instantaneous information on battery charge. This is a difficult assignment for the nickel-cadmium battery because of its constant voltage over most of its discharge cycle. The electrolyte has a high concentration of potassium hydroxide.

There is a great emphasis on battery development, and it may well be that the ultimate battery for wind/solar power sources is being developed in some university or industrial research department at this very moment. Two promising available types are the silver-zinc and silver-cadmium batteries (Fig. 3-21). Cycle life has been improved, and a substantial reduction made in weight and size per given capacity. As shown in Fig. 3-22, the capacity in watt-hours per cubic inch is substantially better than that of the nickel-cadmium and lead-acid types. The graph of Fig. 3-23 demonstrates the weight advantage in watt-hours per pound for the silver-zinc battery as compared with the nickel-cadmium and lead-acid types. Basic construction of the Yardney Silcad and Silvercel is the same (see Fig. 3-21), except for the electrode makeup. The electrolyte is an alkaline solution.

Discharge characteristics are favorable too (Fig. 3-24). The silver-cadmium type holds a constant voltage for a long discharge time as compared with the other types. The silver-zinc type is interesting in that there is a falloff of voltage during the first two hours of discharge and then a leveling off out to 9 hours. This comparison is based on identical weight sizes. A portable color television camera designed by Philips Broadcast Equipment Corp. uses a silver-zinc 24-volt and 250 watt-hour battery that weighs only $8\frac{1}{4}$ lbs. An Ampex color video tape recorder and camera uses a silver-cadmium 30-volt, 150 watt-hour battery with a weight of only 6 lbs.

Charging is an important consideration, and these batteries should never be overcharged. A constant-current system is recommended. Trickle-charging and float-charging are not recommended. Many of the newer batteries have a deep discharge, but these should never be discharged below the knee of the discharge curve so as to avoid developing unbalanced

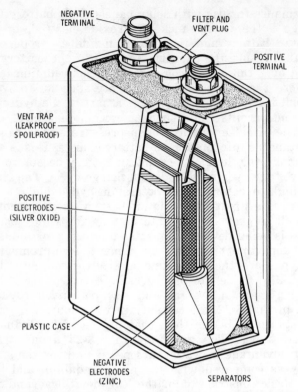

Courtesy Yardney

Fig. 3-21. Basic plan of silver-zinc and silver-cadmium batteries.

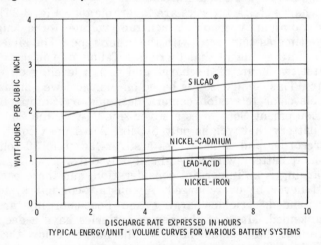

DISCHARGE RATE EXPRESSED IN HOURS
TYPICAL ENERGY/UNIT - VOLUME CURVES FOR VARIOUS BATTERY SYSTEMS

Fig. 3-22. Size comparisons.

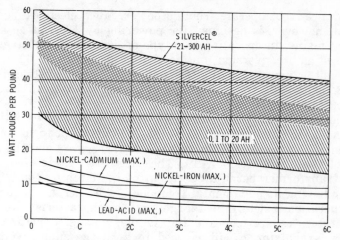

Fig. 3-23. Comparison of Yardney Silvercel®.

cells within the battery. This is also a consideration for the regular nickel-cadmium batteries. The silver-zinc battery has a capability of 100 deep recharge cycles; silver-cadmium, 200 deep recharge cycles.

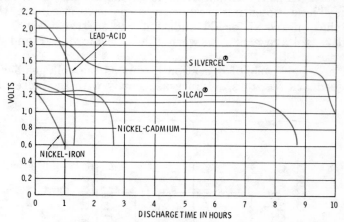

Fig. 3-24. Comparison of discharge characteristics.

INVERTERS

Present solar-cell panel and wind-generator power supplies store their energy in batteries. They are dc power sources. Much modern electrical and electronic equipment is powered from dc sources, such as solid-state electronic units. Direct-current motors are readily available as well as other electrical appliances that operate from dc power. Much electrical equipment, however, is designed for powering by 120-volt ac electrical mains. In operating this type of equipment, an inverter is used as a conversion unit to change dc power to 60-hertz, 120-volt ac power. Usually the conversion is made between 12-volts dc and 120-volts ac, although inverters are available for other input and output voltages.

Most modern inverters are solid state and employ variations of the basic circuit shown in Fig. 3-25. Important to the operation of the circuit are the two power transistors and the saturable base transformer, T1. The saturable characteristic of the transformer permits operation of the circuit as an oscillator having a fundamental frequency of 60 hertz. Frequency of operation is determined largely by the characteristic of the saturable transformer and the supply voltage. Sometimes a voltage-regulator circuit is used to keep the supply voltage constant, ensuring a more constant operating frequency.

With power application, one of the transistors of the matched pair conducts more rapidly than the other. This transistor goes toward saturation and develops one polarity of operation. In the example shown, transistor Q2 at saturation develops positive voltages at the transformer locations marked with the + sign. These potentials result in a positive feedback that switches off transistor Q1 and switches on transistor Q2. A positive voltage exists between the collector of Q1 and the collector of Q2.

A new half cycle of operation begins as the voltages across the feedback resistors, R_F, increase slowly with the increasing

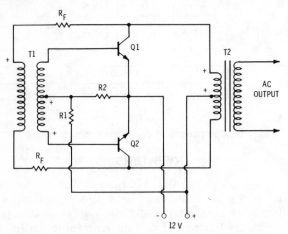

Fig. 3-25. Basic inverter circuit.

magnetizing current in the saturable transformer. At saturation this current increases rapidly, and the voltage across transformer T2 decreases, taking transistor Q2 out of saturation.

Feedback activity now reverses with transistor Q1 going toward saturation and transistor Q2 to cutoff. Now the collector of transistor Q2 is positive relative to the collector of transistor Q1, reversing the cycle across the primary of the output transformer, T2.

Feedback current and saturation of opposite polarity in transformer T2 now return the operation to a new cycle of activity. The transformer ratio of T2 develops a 120-volt ac squarewave across the secondary of transformer T2.

An example of a 500-watt inverter made by Trippe is shown in Fig. 3-26. This very same inverter can also be used as a bat-

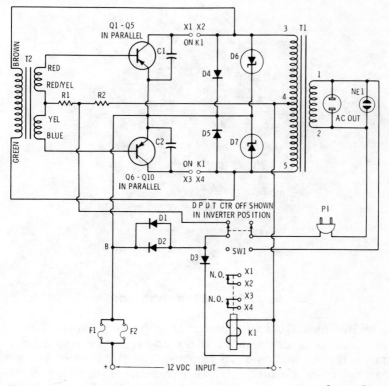

Courtesy Trippe

Fig. 3-26. Trippe 500-watt inverter.

tery charger, which is an attractive advantage for use with solar and wind power sources in case of breakdown or when batteries must be charged additionally because of greater demand. Diodes are used for rectifying and for protective application, and the two zener diodes, D6 and D7, maintain a constant output voltage for inverter operation. A small neon bulb glows whenever there is 110 volts of ac output.

As shown, the control switch is on the inverter position, and the relay contacts close the circuit at X1-X2 and X3-X4. In this position, the bank of transistors (ten of them) is in operation along with saturable transformer T2. Resistors R1 and R2 provide base biasing.

In the charge position, the dpdt switch is in the downward position (see Fig. 3-26). In this position, plug P1 can be inserted into the 110-volt ac power main. Rectified ac voltage across transformer T1 can then be used to charge 12-volt batteries connected across the plus-minus 12-volt dc input.

Courtesy Heath Co.

Fig. 3-27. Heathkit 240-watt inverter.

A Heathkit 175-watt inverter (Fig. 3-27) provides 12 volts of output at a maximum of 16.5 amperes, and 6 volts of output with a maximum of 25 amperes. The circuit uses a single saturable transformer, T1 (see Fig. 3-28), and a tapped primary associated with the collector circuit, to permit either 6- or 12-volt operation.

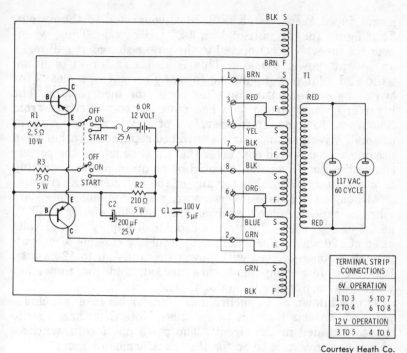

Courtesy Heath Co.

Fig. 3-28. Schematic of Heathkit® Inverter, Model MP-10.

The power output across the secondary is a maximum of 175 watts continuous and 240 watts intermittent when driven from a 12-volt power source. A 6-volt power source makes available an output of 120 watts in both continuous and intermittent modes of operation. Again, the frequency of operation is determined by the dc input voltage and the characteristics of the transformer core. Note that the two windings at the top and bottom of the primary of transformer T1 are connected in the base-emitter circuit of the transistor and are comparable to the secondary of transformer T1, as explained in conjunction with the circuit shown in Fig. 3-25. Resistors R2 and R3 set the small forward bias that permits conduction by the two transistors. Capacitors C1 and C2 minimize voltage transients.

Efficiency of operation is 80%. This tells us that the transfer of power from the dc power source to the ac power output involves a loss of approximately 20%. This is one of the figures that must be considered in planning solar-panel and wind-generator power sources whenever inverters are employed. It must also be noted that an inverter draws a certain amount of power input even though no load is connected to the output. That figure for the unit of Fig. 3-27 is 18 watts. Thus, whenever the inverter is connected to the power source, it will make an 18-watt power demand. This is also a factor in planning wind and solar power sources. In many instances, a means for switching off the entire inverter should be incorporated. The inverter is only turned on when power is to be derived from its output. To go a step further, it is often wiser to use a series of small inverters that can be switched separately rather than use one high-powered converter that must be left on the power-source line continuously. It would be no problem incorporating dc power-source switches for any of the inverters described.

Although solid-state inverters are more common, there are vibrator types available such as the Cornell-Dubilier 12TV12 model of Fig. 3-29. This unit is capable of supplying 115 volts of ac at 120 watts. For example, it would be capable of operating television receivers with power demands up to 120 watts. Relatively few components are needed, and the important vibrator transformer is one of them.

This is known as a synchronous vibrator because it includes pairs of contacts that act as switches. Note that there are no diodes included in the circuit. The pairs operate in synchronism and convert dc to ac for the transformer primary.

The vibrator itself consists of an energizing coil and a vibrating armature-reed assembly that has a specific mechanical vibration frequency. The application of dc voltage energizes the relay coil. It pulls on the armature and closes the contacts which send a burst of current through one section of the transformer primary. At the same time, the relay is de-energized because of the short placed across it by the closed vibrator contact. Thus, it releases the armature, and its natural mechanical motion carries it past the center over to the second set of contacts. When these contacts close, there is a current in the opposite direction through the other segment of the transformer primary. However, the shorting contact across the coil has been broken, and once again the coil is energized and pulls the armature back to the initial set of contacts. This activity continues so long as the dc battery voltage is applied.

Most important, the switching activity causes an alternating opposite-direction current through the primary of the power transformer. The changing magnetic field induces ac voltage across the secondary. Inasmuch as a step-up turns ratio is used, this is a high-voltage ac component.

Topaz Electronics sells inverters with ratings as high as 10,000 volt-amperes; a 3000-volt–ampere model is shown in

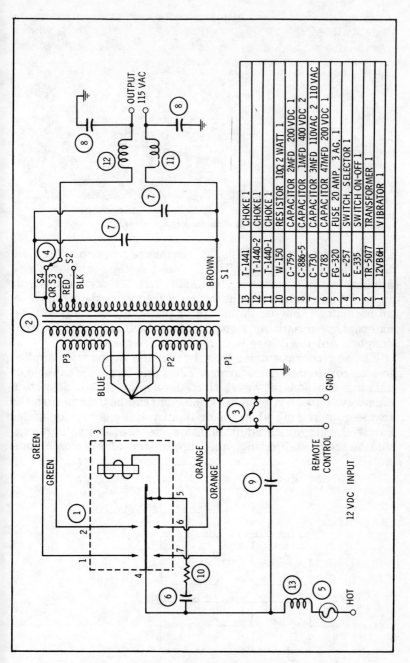

Fig. 3-29. Cornell-Dubilier vibrator-type inverter.

Fig. 3-30. High-powered inverter.

Fig. 3-30. As shown in Fig. 3-31, these are more complex than the lower-powered types. The dc input voltage is filtered and applied to the inverter section. The inverter converts the dc input into a square-wave output at a frequency determined by an oscillator. This oscillator can be placed under control of an external synchronizing signal, making certain that the output frequency of the device is of a specific frequency.

The square-wave drive signal from the inverter is applied to a ferroresonant transformer. This transformer is capable of shaping the square wave, thus developing a low-distortion sine-wave output. A system of detectors can be used to turn the inverter on and off when the dc input voltage is over or under specified voltages. In addition, the model provides voltage regulation, current limiting, and suppression of harmonic com-

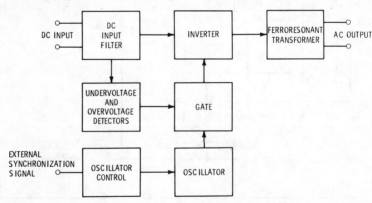

Fig. 3-31. High-powered inverter block diagram.

ponents. Models of various voltage ranges are available and include the common 12- and 24-volt dc inputs and 115- or 230-volt ac outputs.

THE ELECTRIC AUTO

The ideal receptor for energy generated by alternate means is the electric car. In fact, even if you charge its batteries directly from your power mains, you are using energy more efficiently. Comparing an electric vehicle and a standard combustion-engine car and considering all factors—from the sources of the original energy to the completed operational unit—the electric vehicle has a significant advantage. This advantage is further improved if you generate your own electricity with a wind-generator and/or solar-panel power source. One can anticipate that the roof of an electric car will eventually be a solid-state solar panel that can, in the beginning, provide a battery trickle charge and, in later more-advanced designs, also supply a proportion of the main charge for the battery.

Today the electric car has limitations, but these limitations disappear and even change over to advantages in certain applications. As a small runabout to do local shopping in a small town, suburban community, or city district, the electric car has favorable attributes. Cost of operation is several cents per mile, and this does not rise above 5 or 6 cents even when the battery costs are considered. Problems are the weight and capacity of the batteries which limit range. This is just another reason why battery research is rising to a high level.

This is not to say that electric vehicles today are not practical. They are indeed very serviceable if a high-quality version of the common lead-acid storage battery is used. The Sebring Vanguard CitiCar, shown in Fig. 3-32, is a 1250-pound wonder. It is powered by a 3.5-horsepower series-wound 36-volt dc motor. It includes its own battery charger (Fig. 3-33), and one need only run a cable between the car and a convenient 110-volt electrical outlet to charge its batteries. Specifications of the car are given in Chart 3-3.

Note the 38-mph cruising speed and the range up to 50 miles. Within these limitations, the CitiCar is very economical to operate and is an energy-saving means of transportation. Electric costs are estimated at 1 to 2 pennies per mile. Except for charging and care of the batteries, there is little maintenance, and there is a freedom from planned obsolescence. You carry your fuel with you in the form of batteries, and

Fig. 3-32. Sebring-Vanguard CitiCar electric auto.

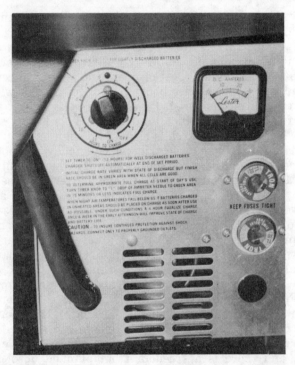

Fig. 3-33. CitiCar battery charger is mounted beneath dashboard.

Chart 3-3. SV-48 Citicar Specifications

Length	95 in
Width	55 in
Wheelbase	63 in
Height	58 in
Front Track	43 in
Rear Track	44 in
Clearance	5½ in
Weight	1250 lb
Rear Storage	12 cu ft
Tires	4.80 × 12 4-ply
Recommended Tire Pressure	35 psi
Speed	38 mph cruising
Range	up to 50 miles
Acceleration	0 to 25—6.2 seconds
Turning Circle	22 ft
Controller	Vanguard multivoltage speed control
Motor	series wound dc 36-V 3.5-hp rating
Power Source	eight 6-volt batteries (HD)
Transaxle	direct gear drive
Suspension	leaf springs solid axles, front and rear; four-wheel shock absorbers on some models
Body	impact-resisting Cycolac (ABS) rust and corrosion proof.
Frame	rectangular aluminum chassis tubular aluminum body support
Brakes	four wheel—front disc rear drum—parking
Gross Vehicle Load Rating	500 lbs

these have to be replaced every two or three years. This is your fuel cost, keeping overall operating costs down in the 4–6¢ per mile range (1975 figures).

As batteries are improved, they can be incorporated into the CitiCar with a resultant increase in range. However, the 50-mile range is quite adequate to meet the needs of many households and small businesses in terms of local-area driving only.

Linear Alpha, Inc. produces their electric autos by modifying Ford Pintos and Mustangs. These are more eye-pleasing and costly but are capable of a cruising speed of 40 mph and a driving range of 70 miles in an 8-hour day. On a single battery charge, the range is 50 miles at a speed of 25 mph. The same company converts Dodge trucks and vans for operation on electric power.

The Battronic Truck Corp. is an autobody works that also manufactures a variety of electric trucks, vans, and other load-carrying vehicles. The small minivan bus of Fig. 3-34 is typical. This 12-passenger jitney bus is ideal for frequent daily trips with intervals set aside for battery recharging. The range of buses, trucks, and vans is difficult to pin down because it is a function of load. Figures fall between 40 and 60 miles. One model in continuous service is capable of 150 stop/starts over a range of 22 miles. This would be excellent for local-area transit buses. Battronics has 16-seat and 22-seat passenger models.

Courtesy Battronic of Boyertown

Fig. 3-34. Small electric bus.

The electric vehicle combines the electric, mechanical, and electronic sciences. Mechanical power is supplied by a dc motor which drives the rear wheels through a direct-gear-drive transaxle. Acceleration is controlled with a speed-control device that is a combination of electronic and electric circuits.

A very basic functional diagram is given in Fig. 3-35. The motor-control block consists of silicon controlled rectifiers or thyristors that regulate the field current of the motor and thus regulate the motor speed and power also. To permit fine control of motor speed and smooth braking, a system of pulse control is employed. The battery voltage applied to the motor is, in fact, controlled by a pulse signal. A large number of pulses (high-frequency pulse rate) result in the application of almost full battery voltage and power. The minimum pulse

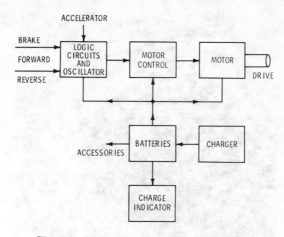

Fig. 3-35. Electric car functional-block plan.

rate sets the effectiveness of the battery power down to about 5%. A smooth control of speed is possible between these extremes. This change in pulse repetition rate is varied with the accelerator of the electric car. Proper activities within the motor and motor-control circuits are established during car braking. Motor operation can be reversed to back up.

Other accessory items, as well as the important charge indicator (Fig. 3-36), operate off the battery. The charge indicator can be in the form of a sensor unit and meter that keep a watch on the battery voltage under load. An alternative plan, shown in Fig. 3-37, is in the form of an electric hydrometer with its sensor mounted within the battery. The meter gives a relative reading of the specific gravity. Some vehicles also include their own charger, and it is only necessary to run a line

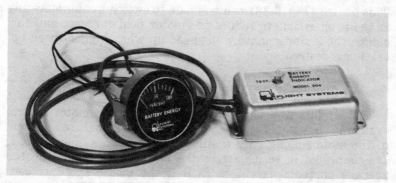

Fig. 3-36. Battery indicator.

Fig. 3-37. Specific gravity charge indicator.

between the vehicle and the power mains to charge the batteries.

Flight Systems makes available a field-weakening system. This device (Fig. 3-38) can be made to switch in automatically at the proper time to reduce the motor field current; and this causes the motor armature to speed up when higher-speed operation is desired.

PRINCIPLE OF THE MOTOR-SPEED CONTROL

The fundamental plan of the Flight Systems motor-control system is shown in Fig. 3-39. Note that the armature and field of the motor, along with a silicon controlled rectifier, SCR-1, are connected in series across the battery. By regulating the average current present in SCR-1, the motor speed is controlled. The silicon controlled rectifier (Fig. 3-40) is a most important device in motor-control systems.

Fig. 3-38. Field-weakening sensor.

The silicon controlled rectifier is a layered npnp device. The end layers serve as cathode and anode, A and C in the symbolic representation. The average current drawn is controlled with the potential applied to the gate element. Inasmuch as a high level of current is drawn, the mounting pedestal also serves as a heat sink for the device. Cathode and gate connections are made at the top.

The SCR or thyristor passes current in only one direction, just as a rectifier diode does. However, the condition of conduction requires that a positive voltage of suitable magnitude exist between terminals A and C (anode and cathode). Furthermore, the gate terminal, G, must be made slightly positive relative to terminal C. Once current is present as a result of a positive voltage being applied to the gate, the SCR current

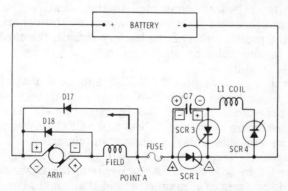

Fig. 3-39. Exide Accumatic motor-control system.

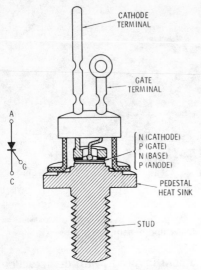

CATHODE
TERMINAL

GATE
TERMINAL

A

G

C

N (CATHODE)
P (GATE)
N (BASE)
P (ANODE)

PEDESTAL
HEAT SINK

STUD

**Fig. 3-40. RCA silicon controlled
rectifier.**

continues to flow even though the voltage is removed from
terminal G. It is apparent, then, that the application of a pulse
to the gate can be used to turn the SCR on.

Current can be cut off by reducing the potential between A
and C or by making C positive with respect to A for a specific
time interval. This, too, can be accomplished with a switched
waveform.

When the device is used in a motor-control circuit, the battery current is drawn in a series of pulses depending upon the
gating of the SCR circuit. This gated current determines the
average current in the motor and therefore the motor speed.

In the actual circuit (see Fig. 3-39), a cycle of operation begins with the application of signal pulse to the gate of SCR-3.
Because its cathode is connected to the negative side of the
battery, and because its anode (by way of the capacitor C7)
reaches the positive side of the battery circuit, the capacitor is
charged to the polarity shown within the circles of C7, as
shown in Fig. 3-39.

A signal is now applied to SCR-1 and SCR-4 at the same
time. As a result, a mean motor current exists because of the
positive charge placed on SCR-1 by capacitor C7. The capacitor will discharge by way of SCR-4 and SCR-1 through coil L1,
placing an opposite polarity charge on capacitor C7 (represented by the charges indicated within the squares beneath
capacitor C7 in Fig. 3-39). Actually, it is the charge held by
coil L1 in the form of a magnetic field that makes certain that
the opposite-polarity charge develops on capacitor C7.

Now another pulse arrives at the gate of SCR-3, and it will again conduct and restore the charges indicated within the circles above capacitor C7. Thus, a new cycle of operation is initiated.

It is understandable that the rate at which the SCRs are switched determines the level of the average current present in the motor circuit. This pulsing of the SCRs is controlled from the logic circuits of the motor-control system. The faster the rate of switching, the higher the average current and the faster the motor speed.

In some motor-control systems, the highest speed of operation is obtained when the motor-control circuit is switched out of operation and a connection is made directly between point A and the negative battery terminal. Also, in a practical system the position of the field winding in the circuit can be reversed and results in a change in direction of the motor rotation.

In an operating electric vehicle, pressing down on the accelerator pedal increases the rate of pulsing of the SCR circuit and results in a higher motor speed and mph. Changing the switch on the control panel between forward and reverse results in a switching of the field winding, determining the direction of rotation of the motor and whether the electric vehicle proceeds forward or backward.

Although the battery current is pulsed, current does flow in the motor circuit even during the intervals when SCR-1 is cut off because of the time constant of the motor field circuit. At these times, diode D-17 provides the complete current path. The speed is regulated by the control that the pulses exert on the average current flow in the motor.

CHAPTER 4

Practical Solar Power
Supplies and Applications

Two factors important in the planning of a solar power supply are average and peak load demands. The peak load demand is closely related to the battery storage system. It must be capable of delivering the peak current demand of the load for whatever period of time such a peak load must be sustained, without a serious loss in voltage regulation. Batteries must be selected in accordance with their 8- or 20-hour rating system. In fact, it is good practice to select a battery or batteries that have a peak current rating (according to their 8- or 20-hour schedule) no lower than the peak current demand of the system.

For example, based on a 20-hour rated battery of 100 ampere-hours, the peak current demand of the system should not exceed 5 amperes ($100 \div 20$). Such a selection means that if your average current demand is lower than the 5 amperes, the storage capacity of your battery will be such that considerably more than 20 hours of operation is feasible. This provides you with backup capacity for a sequence of cloudy days.

The average power demand over a period of time is important in selecting the capacity of your solar panel. This will be brought out in the details of systems covered in this chapter. The importance of the number of hours of sunlight at the mounting site was detailed in Chapter 1. Average charge and discharge information can be estimated and given an adequate safety factor.

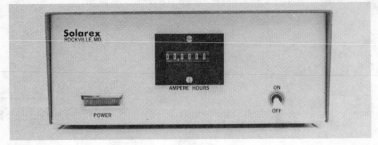

Courtesy Solarex Corp.

Fig. 4-1. Ampere-hour meter.

If a more exact approach is desired, the ampere-hour meter shown in Fig. 4-1 becomes a key instrument. Such a meter can be attached at input and output, and a record can be kept of ampere-hours into the battery storage and ampere-hours out of the storage (Fig. 4-2).

Fig. 4-2. Use of ampere-hour meters.

Ampere-hour readings taken at a specific time each day can give you a wealth of information about the capability of your system and will be of inestimable value in making future expansions or alterations. A difference reading calculated at the same time each day will tell you if the net battery charge has increased or decreased. For example, if the net ampere-hours have increased by 5 ampere-hours on the output and increased by only 2 ampere-hours on the input side, there has been a net battery loss of 3 ampere-hours. By starting out a sequence with the battery fully charged, such daily readings give you a good idea of the battery state. Specific gravity readings can further back up your evaluation of the storage system.

SOLAR-POWERED AMATEUR STATION

The essential units of the solar-powered station operated by W3FQJ are the Argonaut 5-W transceiver, the charging panel, and the 5.5-ampere-hour motorcycle battery, shown in Figs. 4-3 and 4-4. The source of energy is the roof-mounted Spectrolab light-energy converter (Fig. 4-5). This is a 12-volt, 300-milliampere unit.

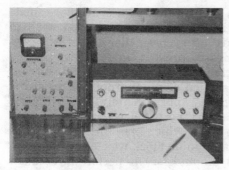

(A) 5-watt transceiver and charging panel.

(B) 5.5 ampere-hour motorcycle battery.

Fig. 4-3. Solar-powered amateur station.

The $5 \times 2\frac{1}{2} \times 5\frac{1}{2}$-inch battery is positioned behind the charging panel for normal operation. Such a battery can be purchased at local motorcycle shops or by mail from one of the auto accessory houses. Mail-order types are shipped with a dry electrolyte which must be added to the battery, along with water, according to the instructions. The electrolyte is already added for you if the battery is purchased at a local shop.

When the solar panel is operating in optimum sunlight, its output is 3.5 watts (12×0.3). At dawn and again at dusk, and during dark, overcast days, the power level is significantly lower. Nevertheless, the installation makes available more than enough electrical energy for a busy 5-watt amateur station.

Twenty hours of accumulative operation at 300 mA produces 6 ampere-hours (0.3×20). For small lead-acid bat-

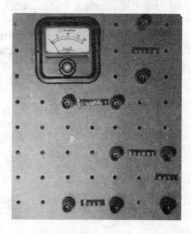

Fig. 4-4. Simple control panel for solar power supply.

teries, this solar panel can be operated as a continuous float-voltage charger, or it can be operated whenever you wish to recharge a battery. In charging even smaller batteries, a resistor can be added in series to prevent an excessive charge rate. Or, the light converter can be used as a low-current trickle charger with the insertion of an appropriate resistor.

The size of the solar panel is approximately 3 × 39 inches. For this particular installation, the panel was mounted on a 5-foot section of television mast which was bracketed to the vent pipe on the roof. Flat aluminum stock was used to fashion a homemade support (Figs. 4-5 and 4-6) to permit adjustment

Fig. 4-5. Roof-mounted solar panel.

of the tilt angle of the panel. A U-bolt permits the top of the panel to be moved up and down the mast. At the bottom of the panel, a flat piece of aluminum with a series of holes permits easy accommodation of practical tilt angles.

The recommended tilt for the panel corresponds approximately to the latitude of the site in degrees north or south of the equator. This particular station is reasonably near the 40° north latitude line. This figure refers to the angle of tilt away from the horizontal, as shown in Fig. 4-7. The nearer you approach the equator, the nearer the optimum mounting angle approaches a horizontal position.

In setting up the mounting arrangement, the solar panel becomes the hypotenuse of a right triangle (Fig. 4-8). Sine and cosine functions can then be used to determine lengths of the vertical and horizontal sides. The overall panel length is

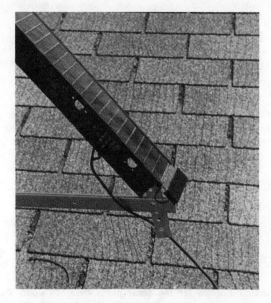

Fig. 4-6. Arrangement for adjusting tilt of panel.

39 inches; this becomes the length of the hypotenuse. There-fore, the horizontal and vertical side dimensions become:

$$b = h \sin A \qquad\qquad a = h \cos A$$
$$= 39 \text{ inches} \times \sin 50° \qquad = 39 \text{ inches} \times \cos 50°$$
$$= 39 \times .7660 \qquad\qquad = 39 \times .6428$$
$$= 30 \text{ inches} \qquad\qquad = 25 \text{ inches}$$

where,

a is the vertical (altitude leg) side,
b is the horizontal side,
h is the hypotenuse (39 inches),
A is the acute angle formed by the hypotenuse and the alti-tude leg.

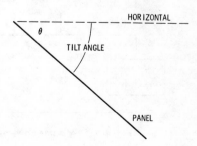

Fig. 4-7. Tilt angle of the solar panel.

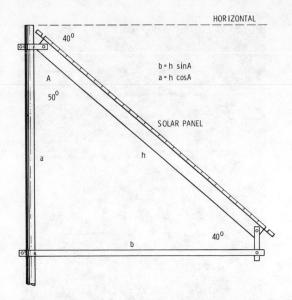

HORIZONTAL

40°

b = h sinA
a = h cosA

A

50°

SOLAR PANEL

a

h

40°

b

Fig. 4-8. Proper tilt angle for the solar panel depends upon the latitude of your station.

The vertical side is an appropriate section of the 1½-inch mast section. The horizontal support is two 1½-inch aluminum strips bolted together but separated where they wrap around the mast section and where they connect to the short length of aluminum strip that permits the adjustment of tilt angle.

Binding posts, meter, protective diode, and hardware are needed for the construction of the charging panel. Refer to the schematic diagram in Fig. 4-9. About 50 feet of two-conductor cable (AWG No. 12 or No. 14 conductors) connects the solar power converter to the charging panel. The panel itself supports a 0–200- or 0–300-dc milliammeter, a 1-ampere silicon diode CR1, and various binding posts and interconnecting wires that permit ease in monitoring and experimentation.

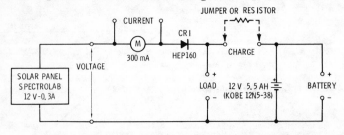

Fig. 4-9. 12-V solar power-supply schematic diagram.

Two binding posts at the top right of the control panel shown in Fig. 4-4 are used to monitor the voltage delivered by the solar panel. The meter to the left reads the actual current being delivered by the solar cells. The meter can be taken out of the circuit by inserting a jumper between the two binding posts labeled "current."

The battery terminals of the control panel are at the lower right. In a continuous-service application, the 12-volt device being powered is connected across these two terminals. When the battery is to be charged, a jumper is connected between the two terminals, labeled "charge" in Figs. 4-4 and 4-9. If you wish to limit the charge to a specified current level under bright conditions, a resistor can be connected between these two terminals.

The two load terminals at the bottom center of the panel permit one to supply energy directly from the solar panel to a 12-volt device. In this connection, the battery is disconnected completely by making certain that there is no jumper or resistor connected across the two charge terminals.

Diode CR1 is an important part of the charging system. It is possible, because of full charge or low illumination, that the battery voltage would exceed the charging source voltage. Without the diode in the circuit, the battery would then discharge into the solar source. This is avoided because, under the condition of high battery voltage and low charge voltage, the diode is reverse biased, the anode becoming more negative than the cathode.

A 5.5-ampere–hour capacity battery provides a conservative and well-regulated source for the 5-watt Argonaut transceiver. Transceiver specifications suggest a 12-volt, 1-ampere source, although peak current demand by the transceiver is less than this value. At any rate, such a battery would supply the unit for many hours of operation. On a 5-watt peak, the current demand would be approximately 417 milliamperes $(5 \div 12)$. In single-sideband operation, the average current demand is substantially lower than this value. Furthermore, the demand current is even less on receive.

The battery itself, based on a 20-hour 5.5-ampere–hour rating, could provide a continuous current demand of:

$$I = \frac{Ah}{T}$$

$$= \frac{5.5}{20}$$

$$= 275 \text{ milliamperes (20 hours)}$$

where,

    I is the current in amperes,
    Ah is the ampere hours,
    T is the time in hours.

This indicates plenty of battery capacity for extended periods of operation. A charging current of the same value would recharge the battery in the same amount of time plus some needed additional time because the charging efficiency is not perfect.

In normal amateur and in most radiocommunication services, deep discharge of the battery and continuous high-current charging of the battery are not necessary. Once the battery is fully charged, a float-charge arrangement is quite adequate. If you assume a charge rate of one-quarter to one-third of the rated current value, you are considering a minimum charge current in the 60- to 90-milliampere range. In the practical installation this level of current, and higher, is readily available for hazy to medium-bright days. A generous current (up to 300 milliamperes) is available on bright sunny days. Even on an overcast day, the solar panel supplies current at the low end of the range (around the 60-milliampere figure). Charging current is quite limited for dark days, but the battery capacity is adequate even for very active QRP (low-power amateur transmission) operation.

The solar power converter can be used to charge different types of batteries with similar or lower capacities. A simple series circuit can be used to charge nickel-cadmium batteries. The series current can be regulated with resistors or a potentiometer (see Fig. 3-16) for a standard type of power-mains charger. You can experiment with resistor values to obtain the desired charging current under bright sunlight. Better regulation is required for charging nickel-cadmium batteries than for ordinary lead-acid batteries. Therefore, this particular solar power converter is used to best advantage in charging 9-volt and 6-volt nickel-cadmium types.

The float-voltage (constant-voltage) charge system is preferred for the gelled-electrolyte batteries. The output is modified to charge 9-volt and 6-volt batteries by using zener diodes to maintain a constant voltage when operating from the solar power source (Fig. 4-10). Recall that for these batteries full charge is obtained with 2.4 volts per cell and under float voltage, a value of 2.25 volts per cell. This corresponds to approximately 13.5 volts for a so-called 12-volt gelled-electrolyte battery. Therefore, a 15-volt or higher solar power source is recommended for charging a 12-volt gelled-electrolyte battery.

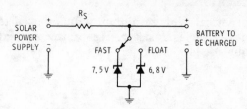

Fig. 4-10. Constant-voltage charging arrangement.

The arrangement of Fig. 4-10 demonstrates how a 6-volt battery can be charged. The actual float voltage required is 2.25 volts per cell, or 6.75 volts (3 × 2.25). A 5-watt zener diode with a voltage rating between 6.75 and 7.5 volts will provide a satisfactory float voltage. At the desired charging current, the ohmic value of resistor R_S would be a value that would provide a drop in the voltage between 12 volts and the zener voltage. If a 6.8-volt zener were used, the value of the resistor for a 100-milliampere charging current would be:

$$R_S = \frac{E}{I}$$

$$= \frac{12 - 6.8}{0.1}$$

$$= \frac{5.2}{0.1}$$

$$= 52 \text{ ohms}$$

where,
  E is the voltage in volts,
  I is the current in amperes,
  R is the resistance in ohms.

A 6.8-volt zener would be satisfactory for float charging, while a 7.5-volt version would do for fast charge. By inserting a switch, you can change over between the two types to obtain fast or float charge, much as the circuit shown previously in Fig. 3-13 does.

VOLTAGE-REGULATED OUTPUT

A regulated output can be obtained from the solar power supply with the use of an appropriate zener diode (Fig. 4-11). A requirement is that the regulated output voltage be less than the voltage of the solar supply. After the voltage of the solar supply drops below a decided minimum voltage, the zener diodes will go out of regulation. The nominal battery voltage is 12.6 volts. A minimum voltage somewhere between 11.5 and

12 volts is practical. This minimum value is used to determine the value of the series resistor, R_S, of the regulated section. The basic equation is:

$$R_S = \frac{V_{S(MIN)} - V_Z}{I_{L(MAX)} + I_{Z(MIN)}}$$

where,

$V_{S(MIN)}$ is the minimum source voltage,
V_Z is the zener breakdown voltage,
$I_{L(MAX)}$ is the maximum load current,
$I_{Z(MIN)}$ is the minimum zener current.

As an example, you may decide to make available a regulated 9 volts. Maximum current demand from such a supply might be 200 mA. Customarily the $I_{Z(MIN)}$ is selected to be 10% of the maximum load current, or 20 mA (0.1×200). By substitution, the value of R_S becomes:

$$R_S = \frac{12 - 9}{0.2 + 0.02}$$
$$= \frac{3}{0.22}$$
$$= 13.6 \text{ ohms}$$

A value of 12 ohms for R is satisfactory. Considering the voltage drop across the resistor and the maximum current, a 5-watt resistor would be more than adequate.

The final step is to determine the power rating of the zener diode. When there is no load on the output of the regulated supply, the total current of 220 milliamperes is present in the diode and the power value becomes:

$$P_Z = V_Z I_Z$$
$$= 9 \times 0.22$$
$$= 1.98 \text{ W}$$

where,

P_Z is the zener power in watts,
V_Z is the zener voltage in volts,
I_Z is the zener current in amperes.

A 5-watt zener diode will perform well in the circuit.

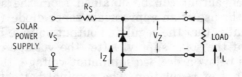

Fig. 4-11. Voltage-regulated power source.

36-VOLT SOLAR SUPPLY

The previous power supply can be expanded into a 36-volt model using the same solar panel. To do so, three 5.5-ampere–hour motorcycle batteries are used instead of one. A jumper arrangement can be used to connect the batteries either in series or in parallel. Hence, in addition to powering a low-powered amateur station, the supply can be used as a solid-state supply for all sorts of solid-state devices using transistors, FETs, and integrated circuits. It has become my bench power supply for solid-state experiments and projects. Depending upon your individual power needs, it may be necessary to add a second 12-volt solar panel. However, solid-state power demands are modest, and the single panel does the job for my needs.

Each battery can be trickle-charged or charged at the 300-mA rate (in bright sun) each third day. Two or three of the batteries can be parallel-charged if desired, or other combined arrangements can be scheduled depending upon your specific needs. Based on a 20-hour discharge rate, the current made available from each battery is:

$$I = \frac{5.5}{20} = 275 \text{ milliamperes}$$

Available battery capacity based on the 20-hour discharge time would be:

$$1 \text{ battery } = 12 \times 0.275 = 3.3 \text{ watts}$$
$$2 \text{ batteries} = 24 \times 0.275 = 6.6 \text{ watts}$$
$$3 \text{ batteries} = 36 \times 0.275 = 9.9 \text{ watts}$$

The above are series connections. In parallel combinations of two and three batteries, power values are:

$$2 \text{ batteries} = 12 \times (2 \times 0.275) = 6.6 \text{ watts}$$
$$3 \text{ batteries} = 12 \times (3 \times 0.275) = 9.9 \text{ watts}$$

The above figures are for conservative operation and are based on 20 hours of demand at 275 mA per battery. How many hours of operation are scheduled weekly? Even more significant, during those operating hours, maximum power demand is intermittent and not continuous. This extends operating-hour capacity.

The 12-V 300-mA solar panel can easily keep batteries up to charge for normal radio operating conditions. In my own case, the solar power supply supports a considerable amount of solid-state experimentation in addition to amateur on-air operation.

Let's take a look at battery capability based on an 8-hour discharge rating:

$$I = \frac{5.5}{8} \cong 700 \text{ mA}$$

Batteries in series:

> 1 battery $= 12 \times 0.7 = 8.4$ watts
> 2 batteries $= 24 \times 0.7 = 16.8$ watts
> 3 batteries $= 36 \times 0.7 = 25.2$ watts

Batteries in parallel:

> 2 batteries $= 12 \times (2 \times 0.7) = 16.8$ watts
> 3 batteries $= 12 \times (3 \times 0.7) = 25.2$ watts

The 8-hour figures indicate that you can draw considerable power from the 3-battery combination and for rather extended periods of operation because of the very nature of the power demand during radio operations. However, a much longer period of time is needed to restore the batteries to full charge because the maximum charging rate is only 300 mA at maximum sun.

WORKING WITH THE SUN

Assume 36 hours of weekly operation at 5 watts. Based on a normal operating situation, a conservative figure for cw (continuous-wave) and ssb (single-sideband) operating time at maximum demand would be one-third of this value, or 12 $(36 \div 3)$ hours.

What would be the weekly watt-hour demand? Watt-hours is a measure of the quantity of electricity and is the product of watts and hours:

$$\text{Watt-hours} = h \times W = 12 \times 5 = 60$$

This amount of energy has to be put back into the battery. In maximum sunlight, the solar energy converter at W3FQJ supplies 3.6 watts (12×0.3). How many hours of a brightness level producing this amount of output would be required to replace the weekly watt-hour demand on the battery, according to the preceding operating schedule? Assuming 100% efficiency, the answer is:

$$h = \frac{\text{watt-hours}}{\text{watts}}$$

$$= \frac{6.0}{3.6}$$

$$= 16\tfrac{2}{3} \text{ hours}$$

This corresponds to approximately $2\tfrac{1}{3}$ hours of saturation-level brightness per day, averaged over a week's time. Indeed, this amount allows very conservative operation of the solar power source. Furthermore, the "one-third" correction factor for normal operation is also conservative, compensating for the fact that the power demand on receive is substantially less than it is on transmit. Even during transmit there is no significant power demand during the spaces of code keying. For single-sideband transmission, the peak demand is made only during the voice peaks, and therefore the average power drawn is substantially lower.

Consider also the fact that there are a great number of hours during which light energy is being converted to electrical watts at a level less than saturation brightness. During overcast and in the early morning and late afternoon hours, watt-hour replacement is being made.

This extra charge time at lower levels more than compensates for the fact that we cannot assume 100% efficiency in the battery-charging process. (There is a great emphasis today in the development of high-efficiency, lightweight, and small-size batteries.)

It is instructive to consider watt-hour capacity in terms of the solar panel itself. Again, let us be conservative and assume that the average brightness is such that saturation current is present for a little more than 3 hours per day. This average load figure more than compensates for dark overcast days and rainy days. On this basis it is safe to assume that there will be approximately 25 hours per week of saturation current. This corresponds to 90 watt-hours (3.6×25). If not maximum, at least significant power is being made available for 3 to 5 hours during the remainder of the day, even during the winter months. Therefore, the weekly capability is near 150 watt-hours. This is an average figure—some weeks more other weeks less. The actual figure depends on geographic location.

A consumption rate of 150 watt-hours per week would operate the system near its limit. This indicates that a continuous demand of 10 watts for 15 hours or 25 watts for 6 hours, averaged over the week, could be feasible. Using the one-third correction factor, this amounts to 45 operating hours at 10 watts, or 18 operating hours at 25 watts.

The above maximum power figures are based on a weekly average. One must be careful not to demand too much power in any one day. For example, if you demand 25 watts for two hours in any one day (six hours of operation), your consumption is 50 watt-hours. Recharge time at maximum sun would be nearly 14 hours ($50 \div 3.6$).

Also, in the same time you will have used a bit more than 4 ampere-hours ($25/12 \times 2$). This is a significant discharge of the 5.5-Ah battery (assuming only a single one was operational). Another hour of such continuous operation without charge would discharge the battery. Continuous and average demand are factors in solar power systems. Each must be considered individually when you are working a system near its limits.

CONTROL PANEL

The control panel shown in Fig. 4-12 has been designed for convenient operation. You may develop plans that better meet your specific needs. However, the description will give you an idea of how the panel can be arranged for versatile charging and use. There are three charge switches that permit the batteries to be charged one at a time or all three simultaneously. This is an aid in dividing the charge current as a function of individual battery use.

A 0–200- or 0–300-dc milliammeter is connected in the charge path to the first battery only. It reads the current delivered from the solar panel when only battery 1 is under charge. If the charge switches are closed to the other batteries, this reading decreases and gives some idea of how current divides among the three batteries. Remember, too, that in parallel connections, batteries charge and discharge each other. Also, a bad battery in the combination can place a heavy load on a good battery. I prefer to charge each one individually.

A pair of binding posts is used for external battery charging directly from the solar panel. An additional pair of binding posts permit a measurement of the solar-panel voltage at the point where the incoming wires connect to the control panel. In operation, batteries can be connected in series or parallel, depending upon the operating voltages to be established.

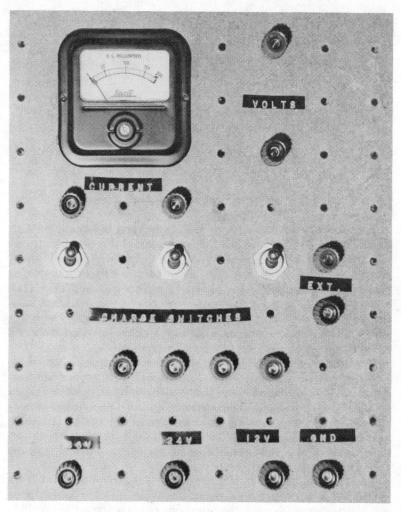

Fig. 4-12. 36-V solar power-supply panel.

The schematic diagram is shown in Fig. 4-13. Note that the negative terminal of the solar panel and the cathode of the protective diode are connected to the poles of the dpst switches. The batteries are connected across the individual terminals of the three switches. Therefore, batteries can be charged separately or at the same time. Battery 1 has its negative side connected to ground. Twelve volts can be obtained by connecting the load between +12 volts and GND. *Whenever 24 or 36 volts is used, switch off all three charging switches.*

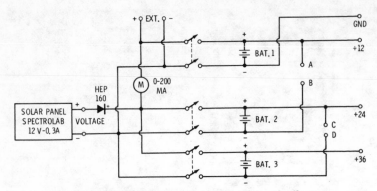

Fig. 4-13. Three-battery control panel schematic diagram.

To obtain 24 volts, a jumper is connected between binding posts A and B. This makes +24 volts available at the output binding post labeled +24 V. When the output is to be obtained from the +36-V binding post, jumpers must be connected between binding posts A and B and between C and D. *Never operate the charging circuit when there are any jumpers connected between A and B or C and D.* A battery short will result if either the battery 2 or battery 3 charge switch is closed with either of these two jumpers in position.

Parallel operation of the three batteries can be obtained by joining the +12-, +24-, and +36-volt binding posts together. Additionally, a ground is established by joining binding posts D and B with ground. The various binding-post interconnections for various series and parallel groupings of the batteries are detailed in Fig. 4-14. The 12-volt single battery operation is inherent. Additional batteries in series or parallel require the use of appropriate jumpers.

The charging of an external battery requires only that it be connected across EXT binding posts. With all charge switches off, the meter will read charging current to the external battery.

INDIVIDUAL CELLS

Silicon solar cells come in a variety of shapes and sizes (Fig. 1-29). These can be assembled in various series and parallel combinations to obtain a desired voltage and current capability. Dimensions can be chosen to meet the requirements of a given enclosure. Cells as large as 3 inches in diameter are available (Fig. 4-15), with current capabilities as high as 500 milliamperes.

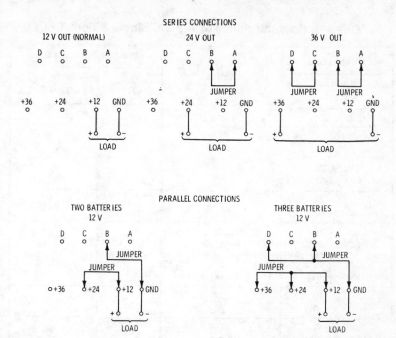

Fig. 4-14. Use of jumpers in making series and parallel battery connections.

Table 4-1 shows cells of various sizes and their output voltage and minimum current characteristics. Usually, cell voltages fall somewhere between 0.4 and 0.5 volt. Hence, quite a group of cells must be wired in series to obtain a reasonable output voltage. Cells in parallel increase the current capability.

Courtesy Solarex Corp.
Fig. 4-15. Large-diameter solar cell.

Table 4-1. Characteristics of Typical Single Silicon Cells

Dimensions Inches (cm) L × W	Active Area in² (cm)²	100 mW/cm² Characteristic	
		Test (1) Voltage (Volts)	Minimum Current @ Test Voltage (mA)
.087 × .20 (.22 × .51)	.013 (.084)	.40	1.7
.10 × .20 (.25 × .51)	.016 (.10)	.40	2.0
.20 × .20 (.51 × .51)	.032 (.21)	.40	4.2
.40 × .20 1.02 × .51)	.065 (.42)	.40	8.4
.80 × .20 (2.03 × .51)	.13 (.84)	.40	17.0
.40 × .40 (1.02 × 1.02)	.14 (.90)	.40	18.0
.40 × .40 (1.02 × 1.02)	.14 (.90)	.43	21.0
.80 × .40 (2.03 × 1.02)	.29 (1.87)	.40	38.0
.80 × .40 (2.03 × 1.02)	.29 (1.87)	.43	44.0
.80 × .80 (2.03 × 2.03)	.61 (3.94)	.40	79.0
1.125 ø (2.86 ø)	.85 (5.48)	.40	110.0

Note that ratings are based on an incident light of 100 mW/cm². This standard value of incident light is often spoken of as equivalent to 1 sun.

How many 1.125-inch-diameter cells would be needed to obtain 6.3 volts at 300 milliamperes? The number of series cells needed to obtain 6.3 volts is:

$$\text{Cell in Series} = \frac{V_T}{V_{Cell}}$$

$$= \frac{6.3}{0.4}$$

$$= 16 \text{ Cells}$$

where,

  V_T is the total voltage in volts,
  V_{Cell} is the cell voltage in volts.

At the 1-sun level, this combination would provide 6.4 volts at
110 mA. To obtain a 300-milliampere capacity, three such
series groups would have to be connected in parallel. Therefore, a total of thirty-eight 1.125-inch-diameter cells would be
needed to obtain 6.4 volts with a capacity of 300 mA.

SOLAR POWER SUPPLIES AT WORK

Modern solar panels and solar power supplies were pio-
neered by the space industry. They are a major source of
power for spacecraft and satellites. Now they serve many
land-based electrical and radiocommunication needs—from
solar-powered golfcarts (Fig. 4-16) and radiocommunication
repeaters to the safety needs of land, marine, and air transpor-
tion.

In Boston a small solar array on the roof of transit cars
maintains a full charge on emergency lighting batteries. The
advantages of solar power installations in many services are
compactness, low maintenance, and elimination of the need to
run electrical lines from sources of power. The primary battery with its need for constant replacement is now supplanted
by a combination of solar panel and rechargeable secondary
battery.

Courtesy Solar Power Corp.

Fig. 4-16. Solar-powered golf cart.

The railroad industry, sometimes slow in accepting new and better techniques, has accepted the solar power supply for various applications (Fig. 4-17). The Southern Railway System operates grade crossings by using solar energy and rechargeable batteries. This combination can also be used in track circuits for switching and for other communication and signal systems.

Courtesy Solar Power Corp.

Fig. 4-17. Railroad application for solar power supply.

Solar-panel power supplies are ideal for inaccessible locations and for areas where it is impossible or too costly to bring in ac lines. A very common place for solar power to be used is at radiocommunication repeater sites. No longer is it necessary to run costly power lines to mountaintops or other isolated locations. No longer is it necessary to be involved with maintenance and replacement of primary batteries. A combination of solar panel and rechargeable batteries can keep a radio repeater operating indefinitely with a minimum of maintenance (Fig. 4-18). A typical repeater may require only 10 Ah per day, which can readily be supplied by a 12-volt, 2- to 3-ampere solar array. A 500-Ah battery would provide plenty of stored energy for a long succession of cloudy days. In fact, such a system is so conservative that there is room for future expansion. Think of the cost of running a power line up to this mountaintop!

The offshore oil platform has become a common user of solar-panel power sources. Such a system can operate lights, horns, and radiocommunications. A major advantage is unattended operations. Sites can be left for long periods of time, and solar energy creates the power that maintains the battery storage.

Fig. 4-18. Mountaintop radio repeater.

Marine applications have become more common, particularly in maintaining a trickle charge when the boat batteries are not being charged by the engine. Thus, the battery charge is kept up when the boat is at dock or when the engine is not operating and the boat is drifting on the water. This application provides an excellent source of emergency power for a sailboat.

One can anticipate the use of the small solar supply on farm vehicles. Many of these remain idle for long periods of time, and battery maintenance has been a continuing problem. Small panels can provide a trickle charge as well as energy for operating accessory items on tractors and other farm machinery.

HEAT AND ELECTRICITY

The great promise for thin-film solar cells is the fact that they can be used both as heat collectors and as generators of electricity. Solar One, shown in Fig. 1-30, is an experimental house using this technique. Research is being done by the University of Delaware Institute of Energy Conversion and Solar Energy Systems, Inc. The latter organization is developing the production technique for cadmium-sulfide thin-film solar cells.

The cadmium-sulfide solar panels generate electricity, which is stored in lead-acid batteries that are housed in the small outbuilding located at the right of Solar One (Fig. 1-30). At the same time the thermal energy transferred through the solar panels is used to heat the air beneath the roof, and this air is directed through a duct system to a storage enclosure housing eutectic salts. These salts store energy in an energy fusion process. By proper selection of eutectic salts, this stored energy can be used for heating in the winter and cooling in the summer.

The direct-current power is used for various dc applications and appliances in the house. Inverters are employed for those electrical devices that must operate on 110-volts ac.

It is estimated that solar energy is capable of supplying more energy than current hydroelectric and nuclear power plants combined—up to 15% of national needs. What is more revealing is that from 20% to nearly 100% of home energy needs could be supplied by solar energy. The latter figure applies when the home is built from the very beginning, using a design that takes full advantage of solar energy. Current types of housing can be modified to derive at least some benefits from the solar energy that is all about us.

A functional plan of the heat and electrical segments of the University of Delaware's Solar One is given in Fig. 4-19. The solar panel/collector combination is mounted on the roof at the proper tilt angle toward the south. Through an airduct system, the heat is drawn away from the solar-energy collectors and guided to the basement. A cross-sectional view of the roof-mounted collector is shown in Fig. 4-20. On top is specially coated Plexiglass which has a long-life transparency. Next comes the combination solar cells and collector. Proper butyl rubber seals and adequate spacers provide a nonleaking, even airflow away from the space warmed by the sun's heat energy.

Additionally, the solar cells convert light energy into electrical energy which is conveyed to the battery house. Here there are twenty 12-volt batteries which are charged. Each has a rating of 80 ampere-hours, and therefore, there is a total capacity of 1600 ampere-hours.

The warm air is processed in the basement by heat reservoirs, one at 70–75°F and the other at 120°F. In between the

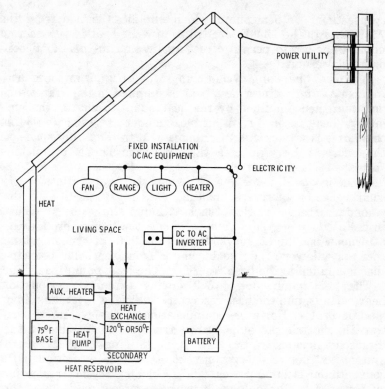

Fig. 4-19. Solar One functional plan.

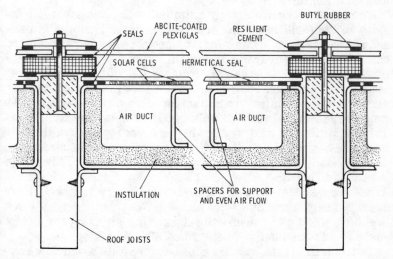

Fig. 4-20. Roof-mounted solar-panel cross section.

two sections is a heat pump, which amplifies the heat from the 75°F level to the 120°F level whenever the heat in the duct is not hot enough to permit direct application to the 120°F secondary reservoir.

A typical operating cycle during a day begins in the morning. On a sunny winter day, heat is supplied to the heat reservoir and then amplified by the heat pump to charge the secondary heat reservoir. A heat exchanger is associated with this latter reservoir and distributes heat to the living space of the house. Solar One is a three-bedroom house, completely furnished.

Later in the day the sun rises and the temperature of the solar collector increases, making a direct connection to the secondary reservoir. Now the heat pump shuts off. The heat in the secondary reservoir now warms the house through the afternoon and night hours. During the late afternoon hours, there is a lower-grade heat coming from the solar collector that is applied to the base reservoir. The heat pump is left off.

The heat pump does work during extended periods of heavy clouds, and the base reservoir is charged by an auxiliary electric heater. However, the operation of the system is such that the electric power is drawn from the mains during the light-load periods of the day. It is anticipated that such a system will derive energy on a basis of 80% solar and 20% conventional electric power. The secondary reservoir also includes a 50°F storage, which in the summertime is used for air conditioning.

The heat reservoirs occupy a limited space and are eutectic-salt storage facilities. Solar heat in the winter is stored in pan-shaped containers containing a special salt named sodium thiosulfate pentahydrate. When the temperature rises to its established level, this salt hydrate melts and dissolves completely in its water. Heat is stored in this liquid. The heat is removed by circulating room-temperature air among the 294 pans, which contain about 700 pounds of salt hydrate. About 750,000 Btu's are held by the thermal storage. In addition to the basic heat-storing salts, there is an agent that is used to induce crystallization as the stored heat is removed. This agent is similar to borax.

As the heat is removed, the salt again crystallizes. However, it keeps recycling with the solar-derived heat, causing the salt to melt and causing the circulating air from the living space to remove energy in the form of heat and so on.

Salts of this type can be processed readily from low-cost and very common natural salts. It is indeed a system of heating that has great long-term possibilities and is nonpolluting. However, it should be emphasized that this is not the only means of using solar energy for heating purposes. However, it is attractive in terms of the availability of the limited roof area on a family-size house, because the solar panel provides a single means of extracting both electrical power and heat energy.

CHAPTER 5

Practical Applications

A recent projection by IEEE's Forecast Group indicates that the electrical power demand for 1985 will be 3.8 trillion kWh, approximately double the 1.85 trillion kWh consumed in 1973. When electrical energy is to be produced in such quantities, one must anticipate that the cost of producing it will rise faster than the rate of consumption. You can assume that your electrical bill for 1985, will be more than double the figure you paid in 1973. Exponentially rising costs are the blight of growing bigness—big business, big unions, and big bureaucracy. No one in the "bigness enclave" is telling us to generate some of our own electricity and, in general, become more self-sufficient as individuals and families. No one in authority is telling us to live in a more decentralized way, with small communities and sections becoming more self-sufficient in many of their needs. Does the demand on big electricity have to double in 12 years? Is it conceivable that you could make your own dwelling or small business self-sufficient in heat and electricity in 12 years?

A small wind generator supplies a small amount of electricity that can be used to supply power for a hobby, such as ham radio, or the outdoor lighting system for your house, etc. A larger wind generator could supply the bulk of the electricity required by a small dwelling. Two modest-size generators can supply more than the usual needs of a family dwelling or farm house. Several large, high, wind generators could supply an entire small town. In all of the previous examples, augmenting power can be supplied by solar panels. It is hoped that efficiencies will rise and costs will decline for these alternate power sources over the next several years.

INSTALLATION OF 200-WATT WIND GENERATOR

The author's own installation for amateur radio station W3FQJ includes two solar panels and a 200-watt wind generator. Specifications are given in Chart 5-1. The site for the wind generator is a small rise, reasonably in the clear, at the

Chart 5-1. Specifications for 200-Watt Winco Wind Generator

Tower	10 feet high
Propeller Type	2 blade
Size	6 feet
Material	Wood
Gear ratio	Direct
Generator	7½″ diameter
	4-pole
Capacity (watts)	200
Approximate maximum amperes	14
Approximate maximum volts	15
Generator speed range (rpm)	270/900
Governor type	22″ Air-brake
Weights	
Generator and parts	61 lb
Tower and propeller	70 lb
Governor assembly	3 lb
Propeller speed range (rpm)	270/900
Wind speed range (mph)	7/23
Average usable KWH Per Month	
10 mph average	20
12 mph average	26
14 mph average	30
Size battery recommended	230 A.H.
Charging Rates:	
270 rpm	0 ampere
350 rpm	2½ amperes
440 rpm	6 amperes
570 rpm	10 amperes
700 rpm	12 amperes
900 rpm	14 amperes

Fig. 5-1. Supporting tower base over wet cement.

back of the house. The Winco Wincharger is supplied with a two-section, 15-foot tower. The base section, shown in Fig. 5-1, has each leg supported by a cement base. Each cement support was constructed by pouring cement into a two-foot-deep hole dug out by a posthole digger.

Sixteen two-foot threaded rods ($\frac{5}{16}$ inch in diameter) were dropped through holes (four holes in each of the base brackets of the tower section) into the cement. The entire assembly was held suspended by building blocks until the cement dried. After drying, the tower section can be dropped down to the level of the cement bases. The *up* and then *down* positions of one of the tower legs are shown in Fig. 5-2. Later, if necessary, the tower can be leveled by using shims under the appropriate bracket. The next step is to assemble the second section of the mast (Fig. 5-3). The top of this section supports the slip-ring assembly.

(A) In position with nuts started.

(B) Tower base dropped down.

Fig. 5-2. Installing base bracket.

As shown in Fig. 5-4, the generator has a fixed field winding and a rotating armature. Whenever the armature is turning, current flows from brush I through field coils D and C and back to brush H. The current is removed by a pair of slip

(A) Top.

(B) Slip-ring assembly.

Fig. 5-3. Second mast section.

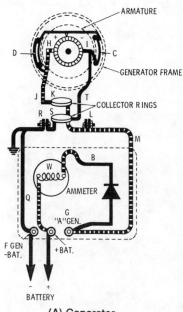
(A) Generator.

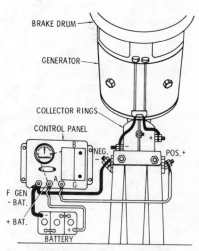
(B) Control box connections.

Fig. 5-4. Wiring diagram.

rings making appropriate connections between the generator and the slip-ring assembly. The slip-ring assembly mounts on a platform; the slip rings and generator revolve as the vane keeps moving the propeller into the wind.

The next step is to attach the generator to the top mast section. Be certain that the generator brackets rest between the two small knobs on top of the collector ring cover. Two ⅝-by-2-inch bolts fasten the brackets to the mounting pipe. A wire pin has been used to hold the turntable in place during shipping. It may be removed after the generator brackets are bolted in place.

This entire assembly must now be placed atop of the lower mast section. A cherry-picking ladder is of assistance in this task. A strong individual can carry the mast and motor assembly up the ladder and hold it in position while others quickly insert four nails, one in each of the four pairs of holes that are used to bolt the top and bottom sections together. These nails will relieve the weight as bolts are inserted and nuts tightened.

The next step is to attach the vane and then the blade. First, though, the governor must be attached to the blade (Fig. 5-5). The governor comes preassembled except for the end plates and the two bolts that will hold it to the brake drum and generator hub.

The bolts must be inserted into the blade holes. This is accomplished by first attaching the end sections. By applying a push-pull pressure to one of the end sections, the springs can be released enough to insert the bolts.

Fig. 5-5. Governor attached to blade.

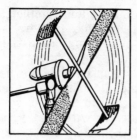

Fig. 5-6. Governor at work.

In normal operating wind, the end plates follow a nonresistive circular path as the blade is rotated in the wind (Fig. 5-6). However, at high speed, centrifugal force opens the end plates and there is a governing action which holds down the blade's rpm whenever the wind speed is greater than 28 mph.

The next step is to attach blade, governor, and brake drum to the generator (Fig. 5-7). This must be done carefully and

Fig. 5-7. Blade attached to generator.

175

evenly. Make certain that the fit is snug and complete, permitting the blade to revolve straight and true. If there is any tilt to the blade after the bolts are pulled up, loosen them again and inspect the components for a true, snug fit.

The bottom of the brake rod, which extends down through the slip-ring assembly, can be folded around into a loop. Another segment of wire can be attached to the loop and brought down to near the bottom of the tower within easy reach. By pulling down on the brake rod, the brake shoe engages the brake drum and prevents rotation of the blade. This step is recommended whenever there is the possibility of winds in excess of 75 mph. However, the wind generator can be shut down any time you wish to minimize wear.

Momentarily short the two wires coming out of the generator by inserting a bolt and nut between their two connecting rings. Tape this to the generator in such a way that wires cannot be broken if the generator assembly is rotated by the vane. It is always a good idea to keep the generator output shorted whenever there is no load placed on its output.

The completed tower and the wind-generator assembly are shown in Fig. 5-8. (Thanks to Richy Atkinson WA3KHM, Bob Bucher WA3KMW, Heinz Frey WA3NDZ, Harry Mullen WA3RLI, and Dick Wagner for their aid and interest in the wind-generator project.)

You can now permit the blade to rotate in the wind to check for smooth operation and freedom from vibration. If all has been assembled properly, there should be no vibration as the blade faces into the wind and rotates. A small amount of vibration is to be anticipated when the wind is gusty and indefinite and the vane is shifting the blade because of changing wind direction.

Wiring the Generator and Control Box

The wiring diagram is given in Fig. 5-4. The first step is to connect the generator wires to the terminals of the collector-ring cover. The two collector-ring terminals can be seen at the top of Fig. 5-3. Small metal tags on the generator wires indicate polarity and determine where the wires are connected to the collector-ring terminals.

There are two terminals on the opposite side of the upper tower support plates. These are the lower pair of terminals shown in Fig. 5-3. From here, wires must be run to the control panel. These wires must be large enough in diameter to carry the current with very little loss. Short lengths and large-diameter wires are preferred. For distances up to 50 feet, use

Fig. 5-8. Completed tower and wind-generator assembly.

AWG No. 6 wire; between 50 and 100 feet use No. 4 wire. Longer distances require correspondingly larger-diameter wire.

The wire from the positive terminal on the tower goes to the A-GEN terminal on the panel (Fig. 5-4B). The wire from the negative terminal on the tower goes to the F-GEN-BAT terminal on the panel. As shown in Fig. 5-4, the A-GEN terminal then connects to the diode and through the ammeter to the + BAT terminal. When the batteries are to be charged, they are connected between the − BAT and + BAT terminals of the control panel.

The wires should be held in place with insulators or suitably taped so that they cannot sway in the wind or be struck by the propeller. In preparing the wire for connection to the control panel terminals, be sure that the insulation is thoroughly scraped off. Bend a loop in the wire before putting it on the terminal bolt; if the wire is bent around the bolt, the insulating washers may be damaged. (Large-ring soldering lugs were preferred at W3FQJ.)

The diode connection is such that the battery current cannot discharge through the generator. In making a preformance check on the generator, it is sometimes necessary to short the diode. This can be accomplished by placing a momentary short circuit between connections B and G (Fig. 5-4).

It is recommended that the charger be grounded against lightning. Connect one piece of AWG No. 4 copper wire to the negative terminal on the tower and the other end to a ¾-inch galvanized water pipe driven 8 feet into the ground. Make good permanent connections at both ends.

In using the wind generator, be certain that good connections are established between the control panel and the battery storage. Also, if batteries are charged in parallel, use appropriate large-diameter interconnecting wires among the batteries. Two 6-volt batteries can be charged in series. Whenever you are charging batteries in series and/or in parallel, it is preferable to use identical batteries in identical states of charge. Parallel batteries do have a tendency to charge-equalize. Clamp-type terminals are preferred, although battery clips can be used if good connections are established.

In the arrangement at W3FQJ, the control panel is mounted directly on the tower (Fig. 5-9). There are very short leads

Fig. 5-9. Control box bolted to mast.

between the tower terminals and the terminals of the control panel. The leads are dressed and taped to the angle iron of the tower. The wind-generator site is more than 500 feet from the operating position of the lead-acid batteries that operate the radio equipment. Consequently, a small wagon (to haul the batteries) and a pair of high-rating batteries are used, with one battery in operating position and the other at the wind generator.

After the generator and control panel have been wired, the circuits can be checked by motoring the generator. This should be done when there is no wind. To motor the generator, release the brake and short a heavy piece of wire between points B and G (Fig. 5-4). This will cause the blade to rotate slowly and the ammeter will indicate a discharge of about 4 to 6 amperes. This will indicate proper wiring of the electrical system.

The wind generator should never be operated open-circuited because high voltages can build up in the generator and damage the commutator and coil. If no batteries are connected, a dummy load or short circuit should be connected across the output. This can be accomplished by disconnecting the minus lead of the battery and then connecting a jumper between F-GEN and A-GEN terminals on the control panel. However, if the generator is not to be operated for a long period, it is wise to put on the brake to minimize wear.

The addition of a solar panel to the wind-generator tower provides a means of obtaining a trickle charge during daylight hours when there is insufficient wind for rotating the blade. The Solarex solar panel is fastened to the south side of the tower and tilted at the optimum angle to make use of the sunlight (Fig. 5-10). Be sure to include a protective diode to prevent battery current and wind-generator output from passing through the solar panel. The negative lead from the solar

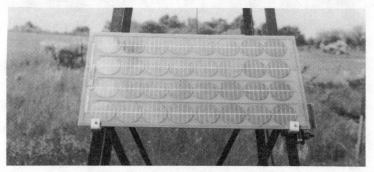

Fig. 5-10. Solar panel attached to mast.

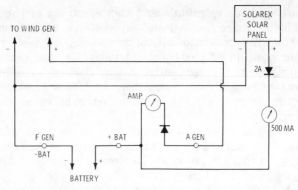

Fig. 5-11. Solar panel connections to wind-generator control system.

panel should be connected to the F-GEN terminal of the control panel; the positive lead to the + BAT terminal (Fig. 5-11). As shown, a 500-mA meter can be inserted.

You may prefer not to make a permanent connection to improve the versatility of the installation. You may wish to have one battery on wind-generator charge while maintaining a second battery on trickle charge by connecting the solar panel to its terminals.

The photograph of Fig. 5-12 shows the Ten-Tec Argonaut and 100-W linear amplifier operating from a lead-acid battery. Solid-state equipment is ideal for solar panel use because of its 12-V dc capability. The combination of wind generator and

Fig. 5-12. Lead-acid battery powers transceiver and 100-W linear amplifier.

solar panels can provide abundant power for operating at least a 200-W peak-envelope-power (PEP) solid-state transmitter/receiver combination.

The operation of 110-volt ac equipment is more cumbersome and requires the use of an inverter (Fig. 5-13). For this installation a 500-W inverter is necessary. In addition, three good-quality lead-acid batteries (connected in parallel) are needed to supply the filament and plate power with good regulation. Maintaining full charge on three batteries places a burden on the solar power supply that cannot always be met when there are successive days of no wind. An auxiliary set of batteries is required.

If you decide on solar power, it is recommended that you change over to 12-volt solid-state equipment, although more power could be made available by a second wind generator of the same size or by going to a 750-W to 1200-W wind generator. The latter generator could supply adequate power for a solid-state 1-kW transceiver.

Fig. 5-13. Transmitter-receiver (bottom) with inverter (top left).

JACOBS WIND GENERATOR

Highly successful wind generators were developed by the Jacobs Wind Electric Company and were produced between the early 1930s and the mid-1950s. Some of these generators are still in operation; others are being reconditioned for operation. Many of the developments and ideas of these early generators are bound to be incorporated in models yet undeveloped for manufacture in the United States. Two of these models had generator output ratings of 2500 watts at 32 volts and 3000 watts at 110 volts. Using a 15-foot-diameter three-blade propeller, they would develop 300 to 400 kilowatt-hours per month. (Over the last year, discounting the electric water heater, the power consumption in my own dwelling averaged less than 300 kWh/month.)

The Jacobs employed a six-pole shunt generator that was direct-driven by the propeller at a speed ranging from 125 to 225 rpm. The rated kilowatt hours were obtained when there were two or three days per week with 10- to 20-mph winds.

The blades were shaped from sitka spruce planks. After shaping and sanding, they were covered with an asphalt-based aluminum paint. Some of these propellers built 25 years ago are still in operation. No brake was employed. Rather, the tail vane was hinged in such a manner that it could be locked straight behind the generator or swung off to the side when the propeller was deactivated.

Jacobs' developments were unique and many are used today. For example, the wind generator employed a fly-ball governor with weights mounted on the hub of the propellers. As a result, the centrifugal force at higher wind speeds twisted all blades a like amount, changing their pitch. Propellers were feathered automatically, and they slowed down in high winds. Generators were built to match the efficiency of the propeller, permitting direct drive. The generator and propeller combination was such that the propeller increased its speed directly with the wind, up to a top speed of 18 to 20 mph.

Electrical systems were such that the generator went into operation at about 125 rpm, reaching full output at 225 rpm. A layered graphite and carbon brush was designed, resulting in minimum wear and a long life of up to ten or fifteen years.

Static discharges and lightning charges often caused arcing in a high wind-generator installation. Heavy grounding brushes and a filter capacitor connected between generator brushes and frame eliminated these problems.

An elaborate control and regulator system was employed. Refer to Fig. 5-14. (Such systems are bound to be solid-state when modern wind generators are developed.) At the top is the generator showing the six fixed field poles and the commutator. The upper and lower brushes on the right side connect to the bottom slip ring and go to the positive terminal in the control cabinet. The two brushes on the left—negative-charging—con-

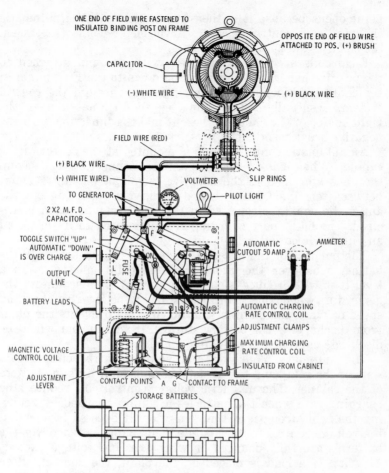

ONE END OF FIELD WIRE FASTENED TO
INSULATED BINDING POST ON FRAME

OPPOSITE END OF FIELD WIRE
ATTACHED TO POS. (+) BRUSH

CAPACITOR

(−) WHITE WIRE

(+) BLACK WIRE

FIELD WIRE (RED)

(+) BLACK WIRE

(−) (WHITE WIRE)

TO GENERATOR

VOLTMETER

PILOT LIGHT

SLIP RINGS

2 X 2 M.F.D.
CAPACITOR

TOGGLE SWITCH "UP"
AUTOMATIC "DOWN"
IS OVER CHARGE

OUTPUT
LINE

BATTERY LEADS

MAGNETIC VOLTAGE
CONTROL COIL

ADJUSTMENT
LEVER

CONTACT POINTS

STORAGE BATTERIES

AUTOMATIC
CUTOUT 50 AMP

AMMETER

AUTOMATIC CHARGING
RATE CONTROL COIL

ADJUSTMENT CLAMPS

MAXIMUM CHARGING
RATE CONTROL COIL

INSULATED FROM CABINET

CONTACT TO FRAME

Fig. 5-14. Jacobs Wind Electric Plant wiring diagram.

nect to the middle collector rings and pass to the middle slip ring to the negative terminal in the control cabinet. The field winding is connected to the top slip ring and connects to the F terminal in the cabinet.

A lead from the positive terminal connects to the main charging fuse and the ammeter and then on to the positive post of the battery system. The negative terminal connects to a large series winding on the cutout relay, to the stationary contact point, and on to the moveable contact point (closed when charging) and the negative battery post.

A pilot light connects between the F terminal and the negative terminal. It glows whenever the voltage-control contact point opens because it is then in parallel with the automatic charging rate control resistor. A voltmeter connects between the positive and negative terminals.

Field coils are shunt-connected, with one end attached to the positive generator brush and on down to the F terminal in the control cabinet. From here it passes through the control system, eventually being connected to terminal 4 near the right center of the cabinet where it makes connection with the negative brush terminal.

An automatic control switch is mounted at the center of the cabinet. When it is off, there is a direct connection made from the generator field to the negative terminal, placing a shunt between terminals 3 and 4, and effectively switching out the automatic control circuit when this mode of operation is desired. The 60-ampere charging fuse is located to its left, and an automatic cutout relay to its right.

The purpose of the latter is to connect the generator to the battery whenever the wind-generator plant speeds up to a level that will produce a charging current. It disconnects the generator when the plant slows down below the charging speed or when the plant is shut down. This prevents the plant from discharging the battery storage. It will cut out when the discharge current falls to 1–1½ amperes.

The automatic charging and voltage-control circuit consists of the relay and two resistance coils shown at the bottom of the cabinet. The magnetic voltage-control coil reduces the charging rate when the batteries are fully charged by reducing the field strength of the generator. It does so by opening the voltage-control contact points. This happens whenever the batteries are fully charged or when the line voltage exceeds 41 volts. If a load is placed on the plant, the contacts close again and produce the normal charging current. It includes an adjustment lever which is set according to charging current and line voltage. The voltage reading is used as a guide to correct adjustment.

When the voltage-control contact points open, the automatic charging rate resistance-control coil is inserted into the field circuit to lower the field current. It also includes an adjustment that determines the level of the trickle-charge current. (The control is set somewhere between 4 and 14 amperes, depending upon the battery capacity.)

The second resistance coil regulates the maximum charging rate. It also is connected as a series resistance in the generator field circuit. However, it controls the top, or maximum, charging rate of the plant. An adjustment clamp permits proper setting. For example, the maximum may be 15 to 20 amperes for a 170-Ah battery storage, while this figure could be increased to 25–30 amperes for a 300-Ah battery capacity.

In effect, the charging rate of the plant is controlled automatically by strengthening or weakening the generator field, giving the battery a proper charge. The control system also provides plant output when the battery is fully charged and is on trickle. Therefore, the wind-generator plant itself will carry the load, and there will be only a minimum battery drain. Meanwhile, the batteries are being maintained at full charge by the trickle current. These plants were built to last; it is hoped that the same objective will be foremost in the modern wind generators under design.

Fig. 5-15. Installation of Del Schlumpberger, KØDEJ.

Del Schlumpberger, KØDEJ, has been operating a reconditioned Jacobs 32-volt, 1800-watt wind plant (Fig. 5-15) for the past ten years. He uses it to operate a variety of power tools and electrical equipment in his shop as well as using it for backup power for the house. The generator is mounted on a 60-foot tower with the battery storage and control point located in a small shed near the base of the tower. He uses 16 lead-acid batteries with a total capacity of 533 ampere-hours.

PRACTICAL S-ROTOR WIND GENERATOR

Michael Hackleman and the group at Earthmind are the developers of a practical S-rotor wind spinner (Fig. 5-16). This particular one is constructed of aluminum impellers (9 feet by 3 feet) and generates 500 watts in a wind speed of 25

Courtesy Michael A. Hackleman

Fig. 5-16. S-rotor Windspinner.

mph. This group is also responsible for developing techniques that use the auto alternator in generating electricity from aeroturbines. Details can be found in Mike Hackleman's book *Wind and Wind Spinners*.

Other wind generators at the site are a reconditioned 1800-watt Jacobs and a 1500-watt Wincharger. The former is a three-blade 32-volt model. The latter is a 32-volt model using a 4-blade propeller. Control panels for the Wincharger are shown in Fig. 5-17. At the very top are the leads coming in

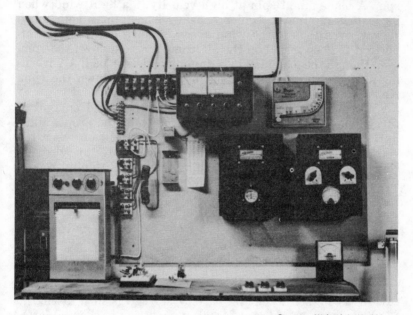

Courtesy Michael A. Hackleman

Fig. 5-17. Wind generating plant control panel.

from the S-rotor, along with the controls and metering for this wind spinner. To the right is a Dwyer wind meter. The two control panels at the middle and left are those belonging to the Wincharger. Battery group selection is made with the array of switches at the left center. Also shown is a strip-chart recorder used for data acquisition during long-term experiments. Battery banks are located to the left of the control setup.

In using an auto alternator and regulator, certain adaptations must be made to meet the needs of a wind plant. When the wind is insufficient, the field current to the alternator must be deactivated to prevent battery drain. In excessive winds, the spinner, alternator, and batteries must be protected. Under operating wind conditions, the alternator current and voltage must be controlled to provide the proper values for correct battery charging.

Loading is an important consideration. When the batteries are fully charged, they place a light load on the alternator and, in turn, a light load on the wind spinner. Under this condition, the rpm must not rise above a safe value. It is a fact that the greater the load placed on the wind spinner, the slower its rpm. A blade might spin at high velocity in a light wind when its load is minimum. Under a heavy load it may spin much slower, even though the wind speed is much higher. It is important to keep a load on the alternator to hold down the rpm, even though the load may be a dummy resistive load. A governing system is a mechanical means of holding down the rpm. No such device is used with the S-rotor.

An alternator generates power only when field current is turned on. This current must be turned on at the proper time. Specifically for a wind plant it must be turned on when the rpm is high enough that the alternator generates an adequate charging current. One way of doing this at Earthmind was to use a wind vane as a sensor of wind speed. Knowing what wind speed produces the correct alternator output permits the sensing device and an associated microswitch adjustment to switch on the regulator and the alternator field current. You do the same thing when you turn on the ignition switch of your auto.

The regulator associated with the alternator used in a wind plant provides voltage and current control. The average field current is higher when the battery storage is in a discharge condition because of the slow switching action of the regulator, keeping the field current *on* for a longer duration than *off*. As the battery storage charges, the average current declines because of the rapid switching action of the regulator.

When completely charged, the batteries draw a minimum current and a light load is placed on the alternator and the spinner. This light loading can be corrected either with a governing system, if that is a part of the wind spinner, or with an electrical control system that increases the load on the alternator. The illustration of Fig. 5-18 is a plan suggested by Earthmind. A second wind-speed rpm sensor and relay are used. If the rpm is too fast because of the light load of the charged batteries, there will be an automatic switchover to a second set of batteries in a discharged condition or a switchover to a dummy load.

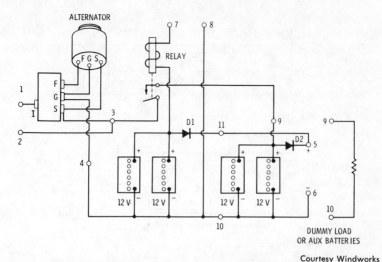

ALTERNATOR

RELAY

DUMMY LOAD
OR AUX BATTERIES

Courtesy Windworks

Fig. 5-18. Automatic load changing when main batteries are charged.

CLEWS WIND-POWERED HOMESTEAD

Henry Clews of Solar Wind Co. has proven in his rural Maine homestead that wind generators are for individuals, families, and small businessmen. He has shown that two wind generators, similar to those shown in Fig. 1-4, can power all the electrical essentials, as well as such items as a color tv, blender, waffle iron, vacuum cleaner, sewing machine, drill press, band saw, grinder, sander, table saw, etc. When wind power is abundant, an automatic relay system helps in heating hot water as well as operating a space heater to supplement the home heating system. All essential lighting and power are handled, including a one-half horsepower deep-well water pump. These include ten 75-watt lights, ten fluorescent lights, a small refrigerator, a radio, a stereo, and miscellaneous appliances.

The initial installation was more conservative and consisted of an Australian Dunlite 2000-watt wind generator. A schematic plan of the system is given in Fig. 5-19. The wind generator uses an appropriate gearing system to drive a 110-volt dc wind generator. This supplies power to a voltage regulator and control panel. Operation of the voltage regulator is similar to that of the auto generator regulators discussed previously. The initial installation consisted of nineteen 6-volt lead-acid batteries that provided 115-volts dc with a capacity of 130 ampere-hours. Direct- and alternating-current outlets are dis-

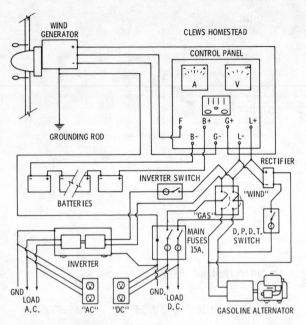

Fig. 5-19. Clews wind-powered homestead schematic diagram.

tributed throughout the house. Most motors and other electrical appliances can be purchased with dc capability. For those that cannot, an appropriate inverter (Fig. 5-19) is used to make the changeover between 115-volts dc and 115-volts ac. A separate power switch is used to turn on the inverter to minimize the battery drain when ac power is not being used. The main power fuse had a 15-ampere rating.

Although a good start, all of the above have been expanded upon to obtain a 60-ampere service with the addition of a 6000-watt Swiss Elektro wind plant. This higher-powered unit with its 16½-foot-diameter propeller is mounted atop a 50-foot guyed tower. Now there are two banks of batteries, and a 3000-watt inverter has been added. It is this combination that provides the almost total service mentioned in the opening paragraph. In fact, the capability of the system is about 600 kilowatt-hours/month.

It is instructive to compare the two wind generators. Both generators are gear driven at a speed about five times faster than the propeller rpm. The blades of both feather and limit the rpm to a safe value in high winds.

As mentioned, the 2000-watt Dunlite uses a voltage regulator. When the batteries are fully charged, the generator output loading is reduced and it is the propeller feathering system that must protect the plant in a high wind. However, there is a manual brake lever that can be set for winds over 75 mph.

The Elektro uses a permanent-magnet alternator, and there is no field current. In a high wind, the tail vane can be adjusted away from the perpendicular and pulled parallel to the blade. Thus, the unit orients the propellers out of the wind. Along with the Elektro, an automatic control can be supplied which functions as a voltage regulator and which also shuts down the generator in high winds. This activity is performed automatically, using a servomotor assembly that shuts the plant down in high wind. It remains so until restarted manually or triggered into operation automatically once every 12 hours. Henry Clews has replaced the time clock to permit triggering a restart every hour.

This Elektro can be operated relatively unattended and has favorable attributes for operation in areas of high wind. The Dunlite requires less maintenance, but must be shut down manually in very high winds.

Table 5-1 was compiled by Henry Clews and appears in his book *Electric Power from the Wind*. Average figures are given for the power requirements of various electrical appliances along with their current requirements for 12 and 115 volts. The fourth column shows the average time that the various appliances are used per month in hours. These figures are then converted to the required kilowatt-hours per month given in the fifth column. Note that the two big items are the electric heat and the electric hot-water heater. There are other better and more efficient means of providing heat and hot water than by electrical means. Electric heat in terms of energy conservation is about the worst method of heating a home.

Chart figures can be added, and a total kilowatt-hours per month obtained for your style of living. As mentioned previously, excluding the energy-wasting hot-water heater, my own dwelling consumes less than 300 kilowatt-hours/month.

Solar Wind Co. is a supplier of 12-volt wind generators, in 200-watt and 750-watt models. They also make available a variety of wind instruments and towers. Enertech Corp. is the agent for Dunlite and Elektro models.

NASA 100-KILOWATT WIND GENERATOR

NASA is experimenting with a 100-kW wind turbine generator at its Plum Brook test site in Ohio. The wind turbine assembly sits atop a 100-foot tower (Fig. 5-20). The rotor is a

Table 5-1 Power, Current, and Monthly Kilowatt-Hour Consumption of Various Appliances

| Appliances | Power in Watts | Current Req'd in Amperes | | Time used per mo. in hrs. | Total KWH per mo |
		at 12 V	at 115 V		
Air Conditioner (window)	1566	130.	13.7	74	116.
Blanket, electric	177	14.5	1.5	73	13.
Blender	350	29.2	3.0	1.5	0.5
Broiler	1436	120.	12.5	6	8.5
Clothes Dryer (electric)	4856	—	42.0	18	86.
Clothes Dryer (gas)	325	27.	2.8	18	6.0
Coffee Pot	894	75.	7.8	10	9.
Dishwasher	1200	100.	10.4	25	30.
Drill—¼ in	250	20.8	2.2	2	.5
Fan (attic)	370	30.8	3.2	65	24.
Freezer (15 cu ft)	340	28.4	3.0	290	100.
Freezer (15 cu ft) frostless	440	36.6	3.8	330	145.
Frying Pan	1196	99.6	10.4	12	15.
Garbage Disposal	445	36.	3.9	6	3.
Heat, electric baseboard, ave. size home	10,000	—	87.	160	1600.
Iron	1088	90.5	9.5	11	12.
Light Bulb, 75-Watt	75	6.25	.65	120	9.
Light Bulb, 40-Watt	40	3.3	.35	120	4.8
Light Bulb, 25-Watt	25	2.1	.22	120	3
Oil Burner, ⅛ hp	250	20.8	2.2	64	16.
Range	12,200	—	106.0	8	98.
Record Player (tube)	150	12.5	1.3	50	7.5
Record Player (solid st.)	60	5.0	.52	50	3.
Refrigerator-Freezer (14 cu ft)	326	27.2	2.8	290	95.
Refrigerator-Freezer (14 cu ft) frostless	615	51.3	5.35	250	152.
Skill Saw	1000	83.5	8.7	6	6.
Sun Lamp	279	23.2	2.4	5.4	1.5
Television (B&W)	237	19.8	2.1	110	25.
Television (color)	332	27.6	2.9	125	42.
Toaster	1146	95.5	10.0	2.6	3.
Typewriter	30	2.5	.26	15	.45
Vacuum Cleaner	630	52.5	5.5	6.4	4.
Washing Machine (auto)	512	42.5	4.5	17.6	9.
Washing Machine (wringer)	275	23.	2.4	15	4.
Water Heater	4474	—	39.	89	400.
Water Pump	460	38.3	4.0	44.	20.

two-blade affair measuring 125 feet in diameter. It consists of
two variable-pitch blades that operate at 40 rpm and permit
the generation of 100 kW of electrical power at 18-mph wind
speed. The mount and drive-train assembly are shown in Fig.
5-21.

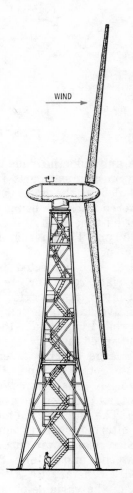

WIND

Fig. 5-20. NASA wind turbine.

Each blade is 62.5 feet long, and performance evaluations
are to be made on all-metal and composite structure types such
as metal and honeycomb. They are designed to deliver 133 kilowatts of power at 18-mph wind speed when rotating at 40 rpm.
Blades are twisted and tapered in accordance with NACA airfoil sections.

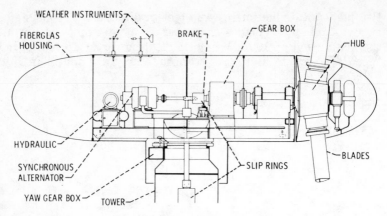

Fig. 5-21. Wind turbine assembly.

A hub links the blade to the low-speed gear shaft. The necessary mechanical parts for changing the blade pitch are also included in the hub construction. Both fixed and articulated teetered hubs are being evaluated. The gearbox is a triple reduction type with a 45:1 ratio, stepping up the rotor rpm to 1800 rpm. The alternator is a synchronous type driven by the high-speed shaft of the gearbox. The alternator is a three-phase, wye-connected 60-hertz model and includes an appropriate regulator. In later experiments the unit will be tied into a power grid, although initial tests are with an all-resistive adjustable load termination. The latter will apply a controlled load under the response of an input signal from a wind-speed sensor.

A large bearing assembly and bedplate permit the entire unit to rotate on top of the tower under the control of a servomotor. A sensor is associated with the servomotor arrangement, and rotates the assembly in accordance with wind direction. The plane of the blades is such that the wind blows over the alternator and assembly into the blades.

Electrical power output will be a function of wind velocity between 8 and 18 mph. But from this wind speed and up to 60 mph, a constant 100-kW output is maintained. This is the result of the variable-pitch blades. Above 60 mph they feather; the same applies for wind speeds below 8 mph.

The wind turbine is turned into the wind for wind direction changes of 10°, provided the wind speed is at least 8 mph for a five-minute interval. The idea is to have the wind turbine respond to broad weather fronts rather than respond erratically to wind bursts and rapid changes in direction.

WIND POWER ITEMS

Ben Wolff and the group at Windworks are noted for their development of honeycomb blades. These strong, lightweight blades, along with a minimum-weight octahedron tower (Fig. 5-22), form an effective wind generator. The systems include a timing belt drive train and slip-ring power transmission. The three blades (12 feet in diameter) can be feathered. The power at the blades in a wind speed of 10 mph is approximately 240 watts. The tip-speed ratio is 5; the design rpm is 117. Samples of honeycomb construction are given in Fig. 5-23. Supplies can be purchased for construction from Windworks.

Jim Sencenbaugh (K6TPS) produces plans as well as a wind-generator kit. Model 750-14 can be constructed by the do-it-yourselfer with elementary tools and about 25 hours of

Fig. 5-22. Wind generator with honeycomb blades.

Fig. 5-23. Honeycomb blade segments.

time. All components are finished and require only bolting together of sections. On the east coast this wind-plant kit is
handled by Solar Wind.

The Sencenbaugh model includes a 3-blade propeller with a
12-foot diameter. It produces 750 watts at 12-volts dc in a 20-
mph wind. In a 10-mph average wind-speed area, you can expect at least 60 kWh per month.

Operation is automatic up to 80 mph because the wind plant
shuts down automatically in winds over 30 mph, restarting
after wind speed has dropped back to 20 mph. The kit includes
a custom-built voltage regulator and a fully wired control
panel with meters and fuses.

Westwind is developing a line of wind and solar power accessories. Two items available currently are a voltage-control
switch and an automatic load-sensing switch.

The voltage-control switch is used with your battery storage set to regulate its voltage, keeping the batteries from
either over or under charging. The device senses high and
low voltage limits. Whenever such a limit is reached, the sensing signal activates a switch that can be used to perform appropriate operations in your wind-generator system. A technique that can be employed is to use the overcharging power to
turn on a hot-water heater, storing the excess energy in the
form of hot water.

The automatic load-sensing switch senses when a load is
turned on and activates a switch. The switching can be used

in a number of ways. For example, it can start an inverter or gas generator, supplying auxiliary power for long periods of no wind. The fact that such a unit can turn on an inverter prevents the inverter from running continuously and placing an unnecessary drain on the battery storage. In this arrangement, the inverter is made to turn on only when an ac device is activated.

Wind power energy need not be stored only as battery power. The water storage system of Fig. 5-24 uses an upper and lower reservoir arrangement. The electric energy from a battery of wind generators is used to operate a pumping system that pumps the water from the lower reservoir to the

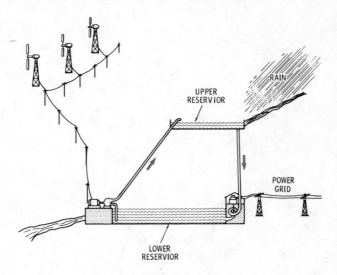

Fig. 5-24. Water storage.

upper one. The force of the water dropping between the upper and lower reservoirs operates a hydroelectric turbine that supplies electric power to the power grid. This is a practical means of using wind generation to add power to a hydroelectric plant.

In fact, electric power companies have tens of thousands of high-tension electric towers feeding electricity all over the country. A modest wind generator mounted on each tower could be a sensible means of augmenting the total power generated by the electric power companies.

APPENDIX

Sources of Supply

BATTERIES

Burgess-Gould
Box 3140
St. Paul, MN 55165

Exide-ESB
Box 5723
Philadelphia, PA 19120

General Electric Battery
1441 NW 6th Street
Gainesville, FL 32601

Globe-Union, Inc.
5757 North Green Bay Avenue
Milwaukee, WI 53201

Gulton Battery Corp.
212 Durham Avenue
Metuchen, NJ 08840

NIFE, Inc.
21 Dixon Ave.
Copiague, NY 11726

Prestolite Battery Div.
Electra Corp.
PO Box 931
Toledo, OH 43694

Union Carbide Corp.
Battery Products Division
270 Park Avenue
New York, NY 10017

Yardney Electric Corp.
83 Mechanic Street
Pawcatuck, CT 02891

ELECTRIC VEHICLES AND COMPONENTS

B and Z Electric Car
1418 West 17th Street
Long Beach, CA 90813

Battronic Truck Corp.
Box 30
Boyertown, PA 19512

Elcar Corp.
2119 By-Pass Road
Elkhart, IN 46514

Electric Vehicle Council
90 Park Avenue
New York, NY 10016

EVS Electric Vehicle Systems
Box 941
Danville, CA 94526

Flight Systems, Inc.
PO Box 25
Mechanicburg, PA 17055

Johns-Manville Sales Corp.
Greenwood Plaza
Denver, CO 80217

Linear Alpha, Inc.
Box 591
Skokie, IL 60076

Sebring Vanguard
US Hwy. 27S Box 1963
Sebring, FL 33870

INVERTERS

Cornell-Dubilier Electronics
118 East Jones Street
Fuquay-Varina, NC 27526

Electric Vehicle Engineering
Box 1
Lexington, MA 02173

Emhiser Rand Industries
7711 Convoy Court
San Diego, CA 92111

Heath-Schlumberger
Benton Harbor, MI 49022

NOVA Electric Mfg.
263 Hillside Avenue
Nutley, NJ 07110

Topaz Electronics
3855 Riffin Road
San Diego, CA 92123

Tripp-Lite Mfg., Co.
133 N. Jefferson St.
Chicago, IL 60606

Wilmore Electronics Co., Inc.
PO Box 2973
West Durham Stn.
Durham, NC 27705

PERIODICALS, BOOKS, AND ASSOCIATIONS

ASE Alternate Sources of Energy
Route 2, Box 90A
Milaca, NM 56353

Electric Vehicle Council
90 Park Avenue
New York, NY 10016

Garden Way Laboratories
Charlotte, VT 05445

Independent Battery Mfg. Assoc.
100 Larchwood Drive
Largo, FL 33540

Mother Earth
Box 70
Hendersonville, NC 28739

NASA
Marshall Space Flight Center
AL 35812

The New Alchemists
Box 432
Woods Hole, MA 02543

New England Solar Energy Assoc.
Box 121
Townshend, VT 05353

Organic Gardening
33 East Minor St.
Emmaus, PA 18049

School of Electrical Engineering
Oklahoma State University
Stillwater, OK 74074

Solar Wind Newsletter
West Allis
WI 53214

Total Environmental Action
Church Hill
Harrisville, NH 03450

Whole Earth Truck Store
558 Santa Cruz Ave.
Menlo Park, CA 94025

Wind Power Digest
Route 2, Box 489
Bristol, IN 46507

SOLAR CELLS AND PANELS

Clairex Electronics
560 S. 3rd Ave.
Mt. Vernon, NY 10550

COMSAT Labs
PO Box 115
Clarksburg, MD 20734

Deko-Labs
Box 12841
Gainesville, FL 32604

International Rectifier
233 Kansas St.
El Segundo, CA 90245

National Semiconductors
331 Cornelia St.
Plattsburgh, NY 12901

OCLI
4501 North Arden Drive
El Monte, GA 91734

Solar Energy Co.
810 18th St. NW
Washington, DC 20006

Solar Energy Systems
70 S. Chapel St.
Newark, DE 19711

Solarex Corp.
1335 Piccard Drive
Rockville, MD 20850

Solar Power Corp.
186 Forbes Road
Braintree, MA 02184

Spectrolab
12484 Gladstone Ave.
Sylmar, CA 91342

TYCO Solar Energy
16 Hickory Drive
Waltham, MA 02154

Vactec Inc.
2423 Northline Industrial Blvd.
Maryland Heights, MO 63043

WEATHER INSTRUMENTS

Danforth Co.
500 Riverside Industrial Pkwy.
Portland, ME 04103

Dwyer Instruments, Inc.
PO Box 373
Michigan City, IN 46360

Kahl Scientific Instrument Corp.
PO Box 1166
EL Cajon, CA 92022

Matrix, Inc.
537 S. 31st Street
Mesa, AZ 85204

Meteorology Research
464 West Woodbury Rd.
Altadena, CA 91001

Solarex Corp.
1335 Piccard Drive
Rockville, MD 20850

Spectrolab
12484 Gladstone Avenue
Sylmar, CA 91342

Taylor Instruments
Sybron Corp.
Box 1, Route 1
Arden, NC 28704

Texas Electronics, Inc.
5529 Redfield Street
Dallas, TX 75209

Wehr Corp.
Climet Instruments Division
1620 W. Colton Avenue
Redlands, CA 92373

Robert E. White Instruments
22 Commercial Wharf
Boston, MA 02110

WIND GENERATORS

Alternative Sources of Energy
Route 2, Box 90A
Milaca, MN 56353

American Wind Turbine
Box 446
St. Cloud, FL 32769

Eastern Sales Co.
1561 Lister Road
Baltimore, MD 21227

Enertech Corp.
PO Box 420
Norwich, VT 05055

Environmental Energies
21243 Grand River
Detroit, MI 48219

Garden Way Laboratories
Charlotte, VT 05445

Michael A. Hackleman Group
Earthmind
26510 Josel Drive
Saugus, CA 91350

Helion
Box 4301
Sylmar, CA 91342

Hexcel
15100 South Valley View
La Mirada, CA 90638

MacDonald Campus
McGill University
College PO
Quebec, Canada

NASA Lewis Research Center
Cleveland, OH 44135

National Research Council
  of Canada
Ottawa, Canada

Pennwalt
213 Hutcheson St.
Houston, TX 77003

Sandia Laboratories
Albuquerque, NM 87115

Sencenbaugh Wind Electric
PO Box 11174
Palo Alto, CA 11174

Solar Wind
Box 7, Bar Harbor Road
East Holden, ME 04429

Standard Research Inc.
PO Box 1291
East Lansing, MI 48823

Virden Perma-Bilt
Box 7160
Amarillo, TX 79109

VITA
3706 Rhode Island Ave.
Mt. Rainier, MD 20822

Walden Foundation
Box 5
El Rito, NM 87530

Westwind
Wilson Mesa
Telluride, CO 81435

Winco-Dyna Technology
PO Box 3263
Sioux City, IA 51102

Windworks
Route 3, Box 329
Mukwonago, WI 53149

Index